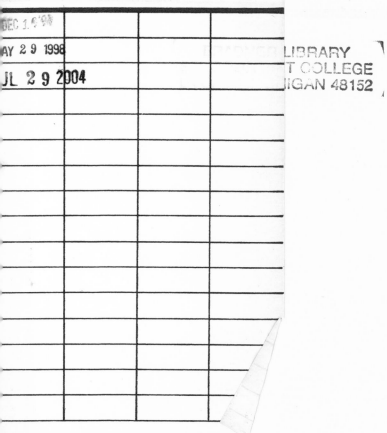

Darwin Machines and
the Nature of Knowledge

HENRY PLOTKIN

═══

DARWIN MACHINES AND THE NATURE OF KNOWLEDGE

HARVARD UNIVERSITY PRESS

CAMBRIDGE, MASSACHUSETTS

1994

Copyright © 1993 by Henry C. Plotkin
The moral right of the author has been asserted

All rights reserved

First published in Great Britain by Penguin Books Ltd

Library of Congress Cataloging-in-Publication Data

Plotkin, H. C. (Henry C.)
   Darwin machines and the nature of knowledge / Henry Plotkin.
     p.   cm.
   Includes bibliographical references and index.
   ISBN 0-674-19280-X
   1. Knowledge, Theory of.   2. Evolution.   3. Evolution (Biology)
I. Title.
BD161.P6 1994   93-39328
121—dc20   CIP

For Victoria, Jessica and Jocelin

Contents

═══

Preface

To know something is to incorporate the thing known into ourselves. Not literally, of course, but the knower is changed by knowledge, and that change represents, even if very indirectly, the thing known. This is ancient folklore but also a commonplace assumption about knowledge that we all make without really thinking about it. It surfaces in various religious and psychodynamic precepts too: know what you fear, for in so doing you take the feared thing into yourself, it becomes a part of your substance, and in that way you control and conquer it. Are these merely primitive folk-tales conjured up by high priests and thaumaturges? Or plain fanciful nonsense? Not at all. There is more than a germ of truth in these claims, which, though they lack the precision of proper analysis, now find support in the science of knowledge. It is just such a science of knowledge that is presented in outline in this book; and, as we shall see, what comes out of it is a smartened-up version of the notion that knowledge is indeed a kind of incorporation of the world, but one of a special sort.

At the heart of the chapters that follow, then, is this proposition that what we commonly call knowledge, be it the name of a friend, the face of an acquaintance or the position of the kitchen in one's house, is a form of incorporating the thing known into the knower. However, it is also something that bears a very close relationship to a much more widespread property of living things, namely the organizational and structural harmony that exists between life and the world in which it has its being. First exposure to this idea leads many to judge it rather weird, and perhaps difficult to understand. It certainly is not an intuitively obvious connection to make. But part of the pleasure of science is the making of unexpected and far-from-obvious links that, when unravelled in greater detail, lead to better understanding of the world. Unfortunately, the fun and fascination of science are often obscured to the uninitiated – that is, to the non-scientist or the not-yet-quite-scientist – by jargon, mathematics and abstract visual depictions.

However, it is not only the arcana of science that the outsider finds impenetrable, but also the insistence of so many scientists in presenting their material to one another in a tone of determined, unrelieved tediousness. Yet the boring appearance of modern science is completely at odds with just how exciting and gripping it really is, particularly in the making of these strange yet profound connections and hence in the unexpected and fascinating ways in which it carves up the world – ways that often seem entirely at variance with ordinary experience.

However odd or difficult the idea that knowledge, human knowledge, what you and I know of the world, is closely connected to a very fundamental feature of all living things might seem at first, I will show that it none the less is important because it tells us significant things about why we ever came to know anything, and how we do so. And what it tells us makes contact with what psychologists and neuroscientists are beginning to understand about our capacity for knowing. In other words, the theory does one of the things that theories ought to do, which is that it makes contact with what scientists of different sorts are thinking and demonstrating. It is also, like so much of science, a very exciting way of thinking, and not actually difficult at all if presented in an approachable way. The object of this book is to do just that – to present this idea, and some that flow from it, in a manner that makes it available to everyone with an interest in how and why we, both as species and as individuals, ever came to know anything about our world and ourselves.

Having written this book for many different kinds of people, some of whom have little knowledge of science, I have tried to keep technical jargon to a minimum and provide a definition when first an unfamiliar word is used. Certain words and phrases, however, appear repeatedly throughout the book (such as 'evolution', 'phenotype', 'heuristic'), and for this reason I have included a glossary that should make unnecessary repeated thumbing back through pages to find those first definitions and explanations, and also serves as a further aid to understanding. The glossary lists the key technical words and phrases that are essential to a science of knowledge in general, as well as to the specific approach of this book. Where my usage is idiosyncratic I give both my meaning and more traditional usage.

The suggested reading at the end of all but the final chapter is meant to serve as a guide to those not familiar with these issues but who find themselves fascinated and want to know more. Some of my readers, though, will be evolutionary biologists, psychologists and neuroscientists of various kinds. These cognoscenti may want to delve deeper into some matters that are not within their areas of expertise, and for them I have provided rather more extended bibliographies for each of Chapters 2 to 6. In order to avoid the visual clutter that chapter-by-chapter reading lists impose, these are presented together at the back of the book.

I am grateful to Celia Heyes, David Hull, Kevin Laland, John Morton and Stephen Walker, all of whom read the original manuscript. Their reactions ranged from encouraging noises to detailed comments and criticisms. The book might have been better had I acted on all of the latter, but ignoring good suggestions is an author's privilege – or vanity. It is certainly the case that the book has benefited greatly from the changes made in the light of their suggestions. I am also grateful to Judith Flanders for her skilful editing.

London
1993

Introduction

The notion of harmony between the organization and structure of individual animals and plants and the world in which they live is almost as old as recorded thought. This is not surprising. The apparent fit, the matching, of living things to the features and conditions of their world – be it the streamlined body shape of animals that move rapidly through water, the camouflaging coloration by which so many animals merge into their surroundings, or the cutting and slashing teeth of carnivores – is a readily observed characteristic of life forms, that immediately impresses itself upon us humans with our special talent for detecting correlated patterns in the world about us. The matching is a result of living creatures somehow incorporating into themselves those aspects of the world that are matched. This is the source of the sense of harmony between the organization of living things and the world about them.

Sometimes the organization is extraordinarily complex, as in the classic example of the human eye; in other cases it may be simple but highly effective, as in the tuning of the hearing receptors of moths to just that frequency of ultrasound used by the bats that prey on them. But simple or complex, match they do. This harmony is the very stuff of so many popular natural-history films and books. It astonishes and delights us, and has been doing so for thousands of years. We call these seemingly clever and often beautiful forms of organization *adaptations*.

Adaptations are the most enduring and powerful focus for attention and study in all of biology. They have been at the centre of metaphysical arguments about harmony and order in the universe, and they have been persuasively used in arguments for the existence of a divine Creator. Prominent in the writings and thoughts of nineteenth-century naturalists, they were a powerful and central spur to the formulation of the theory of evolution that we now call Darwinism or neo-Darwinism. Adaptations are macro-features of the organization of living things and, we now know, are formed by a very long process of interaction between the environment and successions of organisms that make up lineages of organisms extending over thousands, hundreds of thousands and millions of years. They are crucial determinants of whether organisms survive and reproduce or not. It is the study of adaptations that defines biology as different from other natural sciences. And being central to biology they are central also to the human sciences that are not reducible to chemistry; because however special we might think ourselves – and we are indeed special in our unparalleled abilities in just those areas studied by the social sciences – we are, when all is said and done, just a particular form of animal. We human beings are a finely woven cloth of adaptations, as are all other animals.

How, then, can we make that connection between adaptation and knowledge? We do so through a two-track argument. The first is that the human capacity to gain and impart knowledge is itself an adaptation, or a set of adaptations. To the scientifically literate this may not seem to be a startling claim. But it does have specific and interesting implications. We simply will not understand human rationality and intelligence, or human communication and culture, until we understand how these seemingly unnatural attributes are deeply rooted in human biology. They are, I will argue, the special adaptations that make us special. What is unarguable is that they are the products of human evolution, whether adaptations or not. There really are no substantive alternative ways of understanding our extraordinary capacity for knowledge. Creationism of some kind is, of course, another way to look at these things. Perhaps we are able to know what we know because our Creator gave us this ability. But such a view would turn the clock back 140 years. No scientist now accepts creationism as an explanation of anything. Alternatively, there is the curious and indefensible stance of some scientists who study humanity that either we do not need to understand our human qualities in terms of our biology (they somehow just are, and the why and wherefore do not matter!), or if biological recognition is to be given then it must not be in terms of the widely accepted evolutionary principles of selection and descent. None of these are credible, persuasive or interesting positions to take. We are clever animals, and that cleverness needs to be understood, indeed can only be understood, using the same analytical tools and principles that we use to understand our size, shape, gait or metabolism.

The second track of the argument is the one that many find strange and difficult, and one which has already been partially given in the Preface. It is that adaptations are themselves knowledge, themselves forms of 'incorporation' of the world into the structure and organization of living things. Because this seems to misappropriate a word, 'knowledge', with a widely accepted meaning − knowledge usually just being something that only humans have somewhere in their heads − it makes the argument easier if the statement reads 'adaptations are biological knowledge, and knowledge as we commonly understand the word is a special case of biological knowledge'. The line of reasoning that is presented in greater detail throughout the following pages goes like this: the relationship of fit between parts of the organization of an organism, its limb structure for instance, and some feature or features of the world in which it lives, such as the terrain or medium through which it must move, is one in which that organization is in-formed by the environment.

This is the only way to understand the effectiveness of adaptations. If adaptations were formed by mere chance, then the extent to which they work would reflect those same improbable odds. But adaptations, by definition, almost always work. This is because of the in-formed nature of adaptations resulting from the in-forming relationship between that adaptation and its environment. This in-forming relationship between parts of organisms and their world is knowledge, or biological knowledge if one prefers. Now, when a person comes to have knowledge of a particular thing, for example the layout of the keys on the keyboard of a computer (the 'qwerty' keyboard), then the brain state that represents the keyboard is a particular form of organization that also bears a relationship of fit to a feature of the world, the qwerty keyboard, just as the coloration of an insect bears a relationship of fit to the colours of its surroundings. Human knowledge conforms to the relational quality of fit that adaptations have.

So if adaptations are knowledge, and if what we commonly call knowledge (or better, our ability to gain knowledge) is an adaptation, then what in ordinary everyday life we call knowledge is actually a special form of this much wider phenomenon, what I am here calling biological knowledge. Human knowledge is just one kind of a much wider biological knowledge. And science itself is a very special kind of human knowledge. A science of knowledge, then, is a particular kind of knowledge about a special case, human knowledge, that is part of a wider form of knowledge, biological knowledge. If these claims elicit the image of nesting and recursion, of wheels within wheels, that, as we shall see, is entirely appropriate. What follows, then, is the unpacking of this seemingly contorted argument. I will show that our human capacity for knowledge is deeply and ineradicably rooted in human biology; and human biology, of course, is itself rooted in a conceptually and historically wider biology. When we come to know something, we have performed an act that is as biological as when we digest something.

The structure of this book is simple. The first chapter deals with common-sense and philosophical approaches to knowledge and shows that we have now reached the point where a third source of understanding knowledge is possible, and that is through science. I am no philosopher, and this book is not a philosophical tract. But for well over 2,000 years the study of knowledge has been the exclusive property of the philosophers, and they have made important discoveries, the most important and uncomfortable being that human knowledge is never infallible. Now that a science of knowledge is becoming possible it is of real interest to see what light science throws on the age-old philosophical problem of the fallible nature of human knowledge. I do not believe we can write a science of knowledge without at least passing reference to traditional philosophy.

Science, though, is certainly what this book is about, and in my view such a science must be based on evolutionary theory, which is the central theorem of all biology. Chapter 2 presents a non-technical, contemporary view of evolutionary theory. To have merely referred the reader to the many excellent accounts that are readily available would have left the book incomplete and unable to stand on its own as a single introductory source to the science of knowledge. Chapter 3 is a different affair altogether. It departs from standard evolutionary theory and presents the notion of universal Darwinism and 'Darwin machines', that happy phrase of the American neurobiologist William Calvin. In recent years Richard Dawkins has written some splendid things about universal Darwinism. But his message has been the more restricted one that wherever in the universe life has evolved, it has done so by the processes of Darwinian evolution. The version of universal Darwinism adopted here is that in addition to biological evolution as it is normally understood, Darwinian evolution is also operating to produce the transformations in time that we see in certain other spheres, such as immune system function and even the way science itself operates. This is not a new position. It began with Darwin himself and with his friend T.H. Huxley. It has been developed by a number of eminent scholars in the last 140 years. The point of devoting Chapter 3 to universal Darwinism is to prepare the reader for the idea that, in at least some respects, our brains are Darwin machines too, and the way in which we gain knowledge is another form of universal Darwinism.

Chapters 2 and 3 provide the essential conceptual tools for understanding adaptations as knowledge and knowledge as adaptive. Chapters 4 and 5 use these tools to provide that understanding. Chapter 4 considers instinct as unlearned and unthinking behaviour that, none the less, is knowledge in the same way that the camouflaging coloration of an insect constitutes knowledge of its surroundings. Chapter 5 then deals with the reason why a small number of animals have evolved the capacity for altering their behaviour in the light of past experience – that is, why learning and intelligence ever evolved at all.

I am sometimes asked by colleagues and students how the derivation of learning and intelligence, as expounded in Chapters 4 and 5, makes a difference to our understanding of learning and intelligence as we see it now in ourselves and other intelligent animals. One of the answers that I give is that if the arguments of Chapter 5 are correct, and of course I believe that they are, then all animals that can learn and think are born knowing what it is that they must learn and think about. Chapter 6 applies this very important lesson to what we know about knowing in its most obvious and commonly understood form, namely human knowledge. It considers, among other things, the way in which we come to master language, recognize significant people in our lives, reason, react emotionally and share knowledge through culture. It is an explication of what psychologists call domain-specific cognitive function or modularity.

Finally, Chapter 7 returns to philosophy. As already indicated, not only is a science of knowledge now possible, but science has something to tell the classical philosophers of knowledge. And so we return to where we seem to have started. But it is not the same place, because what comes between the first and last chapters is a knowledge of knowledge that traditional philosophy has never had available to it.

A science of knowledge is the first and essential part of a more general project to write a proper science of human beings. We are not able to do that yet, but some time in the future we surely will be. Central to such a science will be a proper understanding of our extraordinary capacity for gaining and communicating knowledge; knowledge that must be understood first as a part of our nature, and only after that as an issue in nurture. Something like the theory presented in the following pages will be a part of that science.

I

The Problem with Knowledge

Knowledge is what gives our lives order. We know who our neighbours are; and which of them can be relied upon for help. We know who we love. We know something of the spatial arrangement of our world so that we can get to and from work, school, shops and our friends' houses. And we know how to manipulate objects: shoelaces get tied, words get written and meals get cooked. We also use knowledge of language to communicate with one another. Without knowledge we could not live, for our world would seem too disordered and we would lose the stable psychological framework that is indispensable to our survival.

This notion, that knowledge is not merely what we know but an indispensable part of our lives, is essential to an understanding of that knowledge. However, if we are really to understand the nature of knowledge, then we are going to have to delve much deeper into the nature of all living things, go beyond an understanding of knowledge of what is or can be known in ordinary, common-sense human terms, and realize that knowing *is* living and surviving. It is central to our lives, and not just our lives but to all life. When we have done that, when we realize how human knowledge is related to other fundamental biological phenomena, and when we have an appreciation of what knowledge is in any living creature, then we will have something approaching a correct understanding of human knowledge, which is an extension of the order of all of life.

These very grand claims for what is usually thought of as a rather ordinary if almost exclusively human characteristic come collectively under the rather indigestible label of evolutionary epistemology. Epistemology is a branch of philosophy concerned with the validity of knowledge – it deals with such questions as how we can know anything, and how we can be certain that what we know is true. Evolutionary epistemology is, in simple terms, the biological study of knowledge. More specifically, it is the study and understanding of knowledge through the use of evolutionary theory. The phrase 'evolutionary epistemology' was first coined in 1974 by Donald T. Campbell, an American psychologist. It is a rather wider and more inclusive phrase than the related 'genetic epistemology' of Jean Piaget, the Swiss philosopher and psychologist. As we will see in a later chapter, genetic epistemology is actually a variant form of the more inclusive evolutionary epistemology. For that reason, when we have to use a wide label, it will be the latter.

'Evolutionary epistemology' is a regrettable phrase because it is both pompous and portentous – it threatens people with an intellectual mugging by a philosopher. However, it does have currency within cognitive science and philosophy. More importantly, the phrase is an explicit reminder that a science of knowledge must be grounded in evolutionary theory. It informs us that knowledge is a problem in evolutionary biology. So it does important work, and we will keep it and use it, if sparingly.

Before developing these ideas about the biological nature of knowledge, which all subsequent chapters will be concerned with, I want first to review some common-sense approaches to knowledge, and to contrast these with philosophical and scientific views. The reason for doing this lies in the fact of our ordinary understanding of knowledge being so different from both scientific and philosophical approaches to the problem, and in the philosopher's approach to knowledge in turn being so different from that of the scientist. If one is to understand what a scientist thinks knowledge is, and most non-scientists think that scientific claims about knowledge are exceedingly strange, then one must realize that scientists themselves appreciate the oddity of their position. This is best done by this scientist, the author, making it clear that he does have an appreciation of what the man or woman in the street thinks knowledge is, and also what philosophers think it is – or indeed, in the philosopher's case, whether knowledge is ever possible. In this way the reader will understand that an evolutionary epistemological view of knowledge is not plain mud-dled thinking but a deliberate move away from both common sense and philosophy. But we will also see that in certain respects a scientific approach to knowledge is much more in sympathy with common sense than with philosophy; and that there are certain common-sense ideas about knowledge that both resonate well with a scientist's view and bear some resemblance to those in one of the great divisions between possible kinds of knowledge that philosophers have drawn.

In which case, why talk at all about what philosophy has to say about the problem of knowledge? The reason is simply that for at least 2,500 years scholars have approached knowledge almost exclusively via philosophy, not science, and no book on the subject can fail to recognize that debt. However, one purpose of this book is to show the reader that science has now reached the point where it too has interesting and significant things to say about knowledge — indeed, that we are coming to the end of philosophy's reign in this area. After millennia of philosophical domination, there is now a serious challenge to epistemology-as-philosophy. The challenger is epistemology-as-science, and one form that it takes is evolutionary epistemology.

SOME COMMON SENSE AND SOME COMMON DISTINCTIONS

At the most general level of understanding, knowledge simply refers to what is known: for example, 'He has great knowledge of farming methods.' Often it is employed in a more collective sense, as in 'Knowledge of farming methods has increased greatly in recent years.' Such usage is fine as far as conversational usage goes. But it is altogether too diffuse and unspecific, and not a little circular, if one's aim is real understanding. In that case, we need to begin with some blend of common sense and precision that will take us within spitting distance of saying what knowledge is in terms that make some connection with science.

The following statement does just that: knowledge, in its most common meaning, denotes a mental state that bears a specific relationship to some feature of the world. For instance, when asked if I know who the only woman Prime Minister of Britain was, I reply that I do know and that it was Mrs Thatcher. It all seems to be such a simple matter that a very important aspect of knowing something is easy to overlook, namely that there are always two aspects to knowledge. The first is a 'feeling' inside the head (how else can one describe the sense of knowing?), or perhaps some other part of the body too, this feeling coinciding with a certain state of the brain. The second is some external event, something outside of the knower's head, to which the feeling relates or refers. When I say that I know that Mrs Thatcher was Prime Minister of Britain, that knowledge claim entails these two conditions: the 'feeling' that leads to the claim, and the person called Mrs Thatcher in the world outside of my head.

Thus knowing is always *about* something, that something being an event or entity in the world apart from our knowing self. We can also know about things that don't actually exist in the material world but do exist in our imagination as well as in books and films: fairies and flying saucers, for example, most certainly do exist in this imaginative, immaterial sense in the world outside of the knowing self, and hence can be known. But we don't and can't ever know in some abstract way such that our knowledge has no object. This *relational* characteristic of knowledge may seem obvious and trivial when spelled out in this way, but it is important because, as I will show, it allows us to consider our ability to know as part of a much wider biological process, namely the formation of adaptations.

Given the relational nature of knowledge, there is an interesting distinction to be drawn between different kinds of knowledge on the basis of the extent to which the internal 'feeling' component includes emotion. There are many instances of knowledge in the ordinary sense of that word which are entirely neutral in that there is no emotional content to the knowledge. The numbers of our houses, the makes of our wrist-watches, the date of the Battle of Hastings or the Gettysburg address and the height of

Mount Everest are all events or states of the world that elicit only the 'feeling' that we know them, and, when necessary, appropriate verbal output such as the year 1066 or the height 8,843 metres. Usually the 'feeling' is so slight that we hardly notice it. Only in tip-of-the-tongue experiences is it sufficiently magnified for us to be really conscious of this feeling of knowing.

Other events, however, elicit not only the 'feeling' that we know them but an accompanying emotional sense as well. Traumatic or frightening events, or the memory of events that are positively highly charged, and all other kinds of emotionally laden happenings, situations and things elicit a range of bodily responses, from localized effects such as 'butterflies in the stomach' and a feeling of constriction of the chest, to the diffuse responses that accompany anger or frustration. Examples are legion and would include events associated with physical pleasure or injury, hostile interactions with people and enduring relationships like friendship and family associations. When we say that we know such an event, the relational characteristic that defines all knowledge is in *this* case made up, at least in part, by an internal 'feeling' about some event in the world which is the product of physiological changes that signal emotional states. So one of the ways that we know someone may, for example, be in terms of our love for them, which is signalled by the bodily states that the knowledge of that person arouse. This is different from, say, knowing the name of the British royal family, in virtue of this difference in the internal component of the relational character that is a part of all knowledge.

The distinction between types of knowledge on the basis of whether it is emotionally laden or not is very much a matter of common sense and ordinary experience. It is, though, important, and we will return to it later.

There is another distinction available to common sense and experience that is equally important, in this case because it taps into a difference that is relevant to both philosophy and the science of knowledge. Interestingly, we can easily gain access to it through the *Oxford English Dictionary*, which contains an impressively large array of definitions of the word 'know', and draws a fundamental etymological distinction between its various meanings right at the beginning. Etymology is the study of the origins and history of words (itself, incidentally, a special form of 'knowing', of gaining 'knowledge') and the etymological distinction discussed by the *OED* goes back two thousand years to the ancient Greeks, who found it sensible to use the word 'know' with two different meanings. One, which appeared many hundreds of years later with its roots in the Latin *noscere* (and subsequently in the French *connaître*), means 'to know by the senses'. The other, which was expressed by the Latin *scire* and then the French *savoir*, means 'to know by the mind'.

Consider what you would say if asked right now what it is you are looking at: your reply would unhesitatingly be 'a book'. You have it in your hands and in front of your eyes, and have been taught that such an object is a book. Thus you *know* that what you experience when you look at what you are holding at this moment is a book. If the book had been placed in the hands of someone who had never seen one before and not been told what it is, such a person would probably respond to the question with a description, something like 'a not very heavy, flattish object made up of individual leaves that are held together at one edge and on which elaborate patterns are drawn'. Not as economical as the label 'book', but it will do for our purposes, because what the person is telling us is that he or she knows 'by the senses' what kind of object it is.

At this point an objection might be lodged that, when answering 'book' to the question of whether you knew what you had in your hands, although you answered on the basis of what was in your senses, the knowledge was buttressed by information already stored in your brain, namely that 'book' is a word used to reference an object of a certain description. Hence the verbal response was based on more than just sensory information. We will see later that this is a constant problem when knowledge is expressed verbally by humans – verbal labels, after all, are learned. And in less obvious form it applies also to what animals know, and how they are able to communicate that knowledge. So consider another example. On a cloudless day a person looks at the sky and knows that it is blue. In this case we might want a slightly more elaborate apparatus to test for what is known in order to get rid of the complications that come with using language. What we might do is ask our subject to match what he or she knows the colour of the sky to be from looking at it, with a colour chart on which is displayed a range of colours, one of which is blue. Such a test of knowledge would present no problems for anyone with normal vision, and would demonstrate that the person knows the sky is blue, and knows it 'by the senses'. (Some philosophers have argued that colours are not primary forms of knowledge, but that is irrelevant to the common-sense view that we are taking here.)

Now, what if our subject had been colour-blind? The subject cannot know 'by the senses' that the sky is blue because he or she cannot see blue (though, as we will see in a moment, the person can know that the sky is blue in another way). However, what the colour-blind person will know is that the sky is cloudless. Whether it is blueness or cloudlessness that is known, it is known by way of an immediate sensory experience, whatever words may be used to describe that experience. This is knowing in the sense of *noscere*, 'to know by the senses'. It matters not how distorted the senses might be or how inadequate the vocabulary used for the description. This knowledge lies in direct sensory experience.

Philosophers who have tried to understand what sets such sensory knowledge off from other forms of knowledge have used words like 'lively', 'forceful' and 'direct', though I don't believe that such labels are helpful. Dreams, for instance, often have vivid qualities but are not instances of sensory knowledge. The nearest that one can get to characterizing this form of knowing, it seems to me, is that what one knows corresponds to something happening in the world that more or less coincides in time with our knowledge of it. Setting aside the time that it takes for light to travel through space and impinge upon receptors in the eye, and setting aside too the time taken for the consequences for the visual system then being interpreted in some way, when I look at the sky and say that it is blue, I mean that it is blue now.

But if a person were to be asked at night what had been the colour (or state) of the sky in the afternoon, and he or she replied that it had been blue (or cloudless in the case of the colour-blind individual), then what is being said is that the subject knows that the sky was blue at some time in the past. It is not blue now, and so the person's way of knowing this must in some way be different from the knowing that derives from immediate sensory experience. In this case, what is known is known 'by the mind', because the knowledge claim relates to events that happened at a time different from that when the claim is being made. It is not possible to know 'by the senses' what is not currently impinging upon the senses.

To be sure, knowing 'by the mind' involves events in the head that correspond to the activation of memory processes, and these occur at the time that the knowledge claim is made, just as knowing by the senses involves the activation of other processes in the head that are more or less coincident with the knowledge claim. But whatever the similarities and differences in these brain processes, and however closely related they both are in time to the knowledge claim, they are different in that knowing 'by the mind' involves a displacement in time between the knowing and the time at which the known events occurred, that displacement in time being absent from knowing by the senses.

Of course, the additional element of memory, some kind of enduring brain state, must exist in the case of knowing 'by the mind', and is responsible for the bridging of that time gap between when the events occurred and any claim to know about them. As we shall see, memories may undergo considerable changes during that period, and such memories may, if one wished, be considered a constantly present, if unconsciously altering, form of knowing. But that would still not change the essential distinction which is being drawn between the case of knowing in which the *external* events that are the subject of knowing are roughly coincident in time with the knowledge claim, and the case in which they are not. (Whether knowledge is conscious or unconscious is not an issue of importance for this book, and so will not be further pursued here.)

Now consider further our colour-blind subject. Through talking to others or reading books, he or she 'knows' that people with normal colour vision see the cloudless daytime sky as blue, and may even 'know' the explanation in terms of physics and the differential scattering of sunlight by the Earth's atmosphere. In this way – that is, by a process of thought – our colour-blind friend 'knows by the mind' that the sky is blue under certain circumstances, even though he or she can never have that knowledge by way of knowing 'by the senses'. So we see that knowing 'by the mind', whether it is understood by invoking memory or thought or both – and later we shall see that memory and thought are very difficult to pull apart – leads to knowledge which in both cases is removed in time from the actual or possible experience of what is known.

Other forms of knowing 'by the mind' include more complex inferential processes. For example, Sherlock Holmes 'knew' that his sinister companion in the train compartment was the vengeful murderer he was seeking, because only the murderer had access to the information that Sherlock Holmes, his greatest threat, would be on this train and in this compartment, and both 'knew' trains to be dangerous places where murder is easily committed. Knowing 'by the mind' also includes logic and mathematics (for instance, knowing that 12 multiplied by 8 is 96, or the proof that the square of the hypotenuse of a right-angled triangle is equal to the sum of the squares of the other two sides). It may be that these forms of knowing are consistently different from one another in terms of 'liveliness' and 'forcefulness'; and it may be that all are less 'lively' and 'forceful' than what occurs by direct sensory experience. I have, though, already indicated that I think this an insecure distinction.

There is yet another aspect of the knowledge possessed by our colour-blind subject that needs to be considered. He 'knows' that cloudless skies by day are blue, not because he can experience blueness, but either because he has learned physics and understands that the atmosphere scatters light in a certain way, or, perhaps not being much of a physicist, he none the less can read and has countless times seen in print that cloudless skies during the day are blue. Here we have another instance of ordinary experience illustrating yet another distinction between types of knowledge and ways of knowing. Knowledge can be gained by the experience of others. It can be shared. And when it is shared, the result is cultural knowledge – or simply culture.

Of all of humanity's attributes, this is certainly one of the most extraordinary and significant, because it can extend the knowledge of individuals through experiences which they themselves have never had. For example, very few people have been to the moon. Those who have, recounted their experience for the many of us who have not. In this manner, those people with the interest but not the means of directly fulfilling that interest know something of what it is like to be on the moon. It matters not, for the moment, whether such knowledge sharing takes place directly by word of mouth, or indirectly, as through the printed page or television screen. What is important is that knowledge, at least of certain kinds, can be shared between people. This has obvious and very important implications for the range of knowledge that can be acquired, and for the rate at which this happens. Not all knowledge can necessarily be shared this way, mind you. And few other species can even begin to do what we are able to in this regard. But we humans can be characterized in a very important way as creatures of culture who share with one another knowledge of almost everything, from dress or the preparation of food, through rituals and myths to the arts and sciences.

In Chapter 6 we will deal with cultural knowledge in greater depth. All that is needed here is for us to note that this is another common-sense distinction to be drawn among the different meanings and experiences of knowledge. The cultural vs non-cultural separation is quite independent of the previous distinctions that have been made. This is particularly the case for knowing 'by the senses' and knowing 'by the mind', since cultural knowledge will involve both. But it is independent also of the emotional/non-emotional distinction.

To summarize, knowledge is always something that comes in two parts. There is the 'knower's end' of knowledge, comprising feelings, brain states and, of course, the means of expressing the knowledge; and there is the 'world's end' of knowledge, which is that aspect of the world that is known. All knowledge is a relationship between the knower and the known. It is also a matter of common experience, as well as useful to this book, to make certain distinctions, among the many that can be made, between different forms of knowledge and knowing. One is between knowledge that is emotionally laden and that which is not. The second is between knowledge of events that are temporally coincident with the act of knowing (knowing by the senses) and knowledge of events that are temporally dislocated from the act of knowing (knowing by the mind). The third distinction is between knowledge that is gained by the direct experience of the knower, and knowledge that is gained through the experience of others, that is, shared vs non-shared knowledge.

We will return at various times to all three of these differences in forms of knowledge. First we must contrast scientific and philosophical approaches to knowledge.

PHILOSOPHY OR SCIENCE?

Ontology is that branch of knowledge theory that is concerned with what is 'out there' that can be known; what, if anything, exists independently of us, the knowers. Ontology ranges from the grand speculations and uncertainties of the metaphysics of existence and creation through to interpretations of how the physical and biological sciences inform ontological theory. Most philosophers and all scientists believe that there is something 'out there' that is independent of us, and this most general claim is, of course, compatible with common sense. The ontological exercise, however, is limited by the formidable problems posed by that other branch of knowledge theory, epistemology. And the profound difficulties raised by epistemology do not stop just at ontology but extend, not surprisingly, to common sense, because the questions that it asks, and the answers that it gives, seem to be so much at odds with ordinary experience.

Whereas ontology is concerned with the objects of knowing, epistemology deals with the methods of knowing. It asks whether knowledge of any kind is possible – and, if it is, what we have to do in order to know. Can we be certain of our knowledge? Indeed, is certain knowledge ever possible? And how can we know which knowledge is certain and which is uncertain? As can be seen from these questions, epistemology is a philosophical exercise in analysis and logic which is carried out independently of ontological considerations. It really does not matter at all to the epistemologist what the nature of the knowable world is, because the questions as to how we can ever know anything are independent of the world that is to be known. What matters is how we come to know, and how sure we can be of knowledge in the face of illusions (false impressions, such as mirages), delusions (untrue or unsupported beliefs) or the possibility that all experience is simply some kind of elaborate dream that is not rooted in a real and knowable world. It is worth remembering that many philosophers have pondered upon the meaning of a commonly experienced dream in which the dreamers dream that they pinch themselves to see if they are dreaming.

Philosophers define knowledge as justified true belief. If I say that I know it is raining, then, for this to be a claim of real and certain knowledge, (1) it must be raining, (2) I must believe it to be raining (merely to say that it is, out of whim, and for it to be raining at the time of the whimsy, would not constitute knowledge that it is raining), and (3) I must be justified in having that true belief. By justify, epistemologists mean that the claim must be judged as reasonable rather than not. For example, I might genuinely believe it to be raining, and it is raining, but my belief may be based on what someone else has told me, and that person may be none too reliable. I may even know that my informant is sometimes economical with the truth. 'How do you *know* that it is raining?' I am asked. 'Why,' I answer, 'because so-and-so told me.' 'Well,' say the philosophical judges on this matter, 'it is indeed raining, and you clearly believe it to be so doing, but your informant is unreliable and therefore you are not justified in your claim. You don't really know with any certainty that it is raining.' Or perhaps I am hard of hearing and have mischievously been told that it is hailing. I am not justified in claiming that I know it is raining because it is by chance that my belief, based on a misperception, matches what is happening outside.

It is this third requirement, justification, that has so bedevilled the epistemological enterprise, and it has always been so. Socrates and Plato struggled with this problem thousands of years ago, and in more recent times Descartes, Hume and Kant expended much of their intellectual effort in the attempt to solve it. A glance at any philosophy primer will show the reader that it remains the central issue of modern epistemology. Such is the magnitude of these difficulties that philosophy has not yet arrived at a single answer but has reached instead, very roughly, two kinds of solutions. The first comes under the general heading of rationalism, which argues that true knowledge can only come through intellectual, deductive activity and never by way of the senses. In terms of our earlier distinction, knowledge is only possible through knowing 'by the mind'. The second, referred to generally as empiricism, claims that knowledge is only ever possible through experience, which means that, initially at least, knowing is mostly 'by the senses'. (Philosophers will, with some reason, feel that I am doing great violence to their subject by so brutally reducing its variety in this way. Philosophically they have a point. There are many other positions that philosophers have adopted, including one, pragmatism, which has much in common with the general arguments presented in this book. However, I have a scientific case to argue and don't want to be deflected from it for too long by matters philosophical.)

Among the famous subscribers to rationalism are Plato and Descartes, the former's allegory of the cave being surely the most famous exposition of a rationalist thesis in all of philosophy. In the allegory of the cave that he gave in his dialogue *The Republic*, Plato suggested that our knowledge of the world, as it comes to us through our senses, is equivalent to the shadows of objects cast upon a cave wall. Such knowledge is incomplete and often misleading – truth of this kind is literally 'nothing but the shadows of images'. However, we can be liberated from such partial and uncertain knowledge by throwing off the shackles that bind us with our backs to the light and to the objects that cast their shadows, and turn around and look directly at the light and what it illuminates. This is a painful experience at first, but gradually, with effort and training, one can overcome the glare and come to true knowledge:

He will grow accustomed to the sight of the upper world. At first he will see the shadows best, next the reflections of men and other objects in the water, and then the objects themselves; then he will gaze upon the light of the moon and the stars and the spangled heaven . . . last of all he will be able to see the sun, and not merely reflections of him in the water, but he will see him in his own proper place, and not in another; and he will contemplate him as he is . . . the prison-house is the world of sight, the light of the fire is the sun, and you will not misapprehend me if you interpret the journey upwards to be the ascent of the soul into the intellectual world.

What Plato meant was that, by application to a specific course in reasoning and mathematics, the philosopher-king (and only philosopher-kings are fit to rule over others) can slowly come to the world of forms and knowledge which we all have within us from birth, but which only becomes accessible by careful and long study. Knowledge for Plato is an ideal achieved through thought. What becomes known has always existed, unchanged, and is separate from the experienced world. This is absolutely certain knowledge. The experienced world, by contrast, can never really be known. And this uncertainty includes the sciences, because the sciences study the experienced world, which is the world of the shadows in the cave.

Rationalist theories of knowledge can be criticized on a number of grounds, not the least being that what constitutes true and certain knowledge varies, often widely, from one rationalist philosopher to the next. This is because the criteria for establishing the difference between true knowledge on the one hand and more ordinary thoughts and reasonings on the other either are given only vaguely, or themselves vary from philosopher to philosopher.

On common-sense grounds too, rationalism seems to be wrong. While conformity to the practical values of the man or woman in the street — which generally are that if something works it is good and, in the case of knowledge, true, whereas if it doesn't work then it is bad and false — is surely not the measure by which to judge a philosophical position, none the less the rationalist notion that ordinary sensory experience is not to be trusted as knowledge does seem to be strangely at odds with ordinary life. No one would like to argue that everyday sensory experiences constitute certain or complete knowledge, but by and large they do seem to work, and this pragmatic value suggests that Plato's assertion about the uncertain and partial nature of sensory experiences is an overstatement. Either our experiences are sufficiently correct — that is, what we know through ordinary sense experience is sufficiently close to true knowledge as to allow our ordinary expectations to be fulfilled most of the time, or we are all of us either exceptionally lucky most of the time or mistaken in believing that we regularly achieve our goals. But the 'lucky' argument is too unlikely to be acceptable. After all, most people do manage to return to their homes successfully, to address appropriately their acquaintances and to place the gear levers of their cars in the right place. The chances of this happening by good fortune even to one person most of the time are vanishingly small. For all of us to be that lucky is just impossible. And the idea of some universal delusional state strikes most people as plain silly. So, at the very least, ordinary experience is based on 'workably' good knowledge. It may not be absolutely certain knowledge, but it is certain enough to work most of the time.

Scientists have never been happy with the rationalist thesis either, in their case because all science is built, in the first instance, on observation. Indeed, scientists go to great lengths to ensure the accuracy of their observations precisely in order to make them more accurate and true. And non-rationalist philosophers too have doubted the notion that knowledge comes to us only by thought and reasoning. Many of the latter have begun with the idea that, as Bertrand Russell put it, 'the body is what brings us into touch with the world of external reality'. And, as might be expected, these philosophers, known as the empiricists, have also been much influenced by the rise of modern science from the seventeenth century onwards.

In fact, the common-sense, scientific and empiricist criticisms of rationalism do have common ground in that the early empiricist philosophers of the sixteenth and seventeenth centuries, like Bacon and Locke, identified themselves with the rise of early modern science, and early science was intensely practical rather than theoretical. Ideas alone were not enough. The ideas had to be based on observation and they had to have practical value, they had to work. Thus the empiricist search for an understanding of knowledge began with the seemingly practical matter of how accurate and certain our sensory information is. The early empiricists also denied the existence of innate knowledge, a point of view that, as we shall see, has extended into this century and was the hallmark of an influential school of twentieth-century psychology, vestiges of which remain with us.

While the empiricist approach to knowledge is rooted in much more recent times than rationalism, it is now hundreds of years old. It has its own long history and has been prominently represented in this century by a school of philosophers known as the logical positivists, or logical empiricists as some prefer to call them. It would take us too far from the science of knowledge to trace empiricism down the last 400 years. So, as with our all-too-brief treatment of rationalism, only one example of empiricist thought will be considered in a very abbreviated form. I have chosen the eighteenth-century philosopher David Hume because of his importance in promulgating scepticism about induction. Induction is a form of ampliative inference in which a conclusion concerning all the members of a class is drawn from direct observation of only some of them. For example, having examined many rabbits and observed that they all have long ears, the inductive inference would be that all rabbits have long ears. It is a kind of probabilistic exercise, and so we might want to examine all the rabbits alive today, and may even have access to information on all rabbits that have ever lived. They all had long ears. So, we infer with even greater certitude that rabbits have always had long ears and always will have long ears. Hume taught us that this form of inference is hazardous. We can never be certain that we will not come across a rabbit tomorrow, next week or next year that does not have long ears. Knowledge from the past does not guarantee the certainty of knowledge in the future. As we shall see, this is a crucial matter in the evolution of a particular form of knowledge, namely that derived from learning and intelligence.

Hume was much impressed by the power of Newtonian thought and its startling success. His starting-point was that the proper understanding of knowledge lies in a science of humanity, specifically a science of mind. Thus Hume was one of the first psychologists, and gave rise to the idea that application of Newtonian scientific methods to psychology will yield a kind of mental mechanics, analogous to physical mechanics, and with similarly powerful results. Like empiricists before him, Hume rejected the notion of innate knowledge, subscribing instead to the maxim that there is 'no idea without an antecedent impression' (sometimes phrased as 'there is nothing in our intellect which has not entered it through our senses'). However, unlike previous empiricists but much in line with his search for mental mechanics, Hume laid great emphasis upon the association of ideas, once these ideas have entered into our minds by our experiences. In particular, his associationism led him to an analysis of our belief in cause–effect relationships which he believed should be understood in the context of our tendency to associate mental events.

There is a great psychological potency, Hume suggested, about our need to attribute a cause–effect structure to the world. He denied the necessary reality behind such a propensity of thought, arguing that our concept of causation was a habit of thought rather than a form of knowledge about the real world. This is because the uniformity of nature, which is the picture that we have of the world as one in which the constant conjunctions of the past and present guarantee their existence in the future, is an inference whose only foundation is our psychological tendency to assume that the experienced pattern of past events (including cause–effect relations) will reoccur in the future. Causation is just a habit of thought, and so too is induction. As an inferential procedure the latter can never be certain. We are psychologically *compelled* to believe that 'instances of which we have no experience must necessarily resemble those of which we have'. Such psychological compulsion does not, though, necessarily reflect the way of the world.

So it was that Hume, who began his analysis with the assumption that only an empiricist-scientific approach to knowledge could yield anything of epistemological value, ended up propounding an extreme form of scepticism about what we can know with certainty. There is no innate knowledge, only our sensory experience and impressions or ideas, and our propensity to see particular patterns in those experiences. However, those patterns do not reflect what is actually in the world, and certainly are no guarantee that the patterns we see today will be seen in the future. In other words, while we have, perhaps, a very limited claim to knowledge of the past and present, we can never know anything of the future. We are not, and never can be, prescient.

Thus, empiricism led to deep scepticism about the possibility of us humans ever having certain knowledge, and all of epistemology since Hume has been an attempt to rescue our understanding of knowledge from Hume's critique of empiricism. Pre-eminent among post-Humean philosophers was Kant, aroused, in his own words, from his 'dogmatic slumbers' by Hume's pessimism. We will return to Kant in the final chapter. Yet despite the serious attention from philosophers such as Kant, empiricism, like rationalism, seems to conflict with ordinary experience. After all, having negotiated a particular route through London's streets each day, I always arrive at a place I know to be my home. This has happened for years past, and I expect, therefore, that it will happen tomorrow as well. Yet the Humean argument about the uncertainty of induction says that I cannot rule out the possibility that something will have happened to my route or the position of my house so that tomorrow I will not get where I want to be.

Hume himself recognized the conflict between ordinary experience (of course I will get home tomorrow) and the logical incompleteness of induction as a basis for certain knowledge (I cannot possibly know with certainty that I will get home tomorrow). So, by and large, the ordinary experience of inductive inference for these kinds of experience or events tells us that induction works perfectly well. Once again common sense seems to win over academic philosophy when it comes to issues like where our houses are in space and whether the sun will rise and set tomorrow. However, common sense and experience also tell us that there are circumstances where Humean pessimism about the adequacy of induction is indeed well justified. Our experience of, say, social relationships tells us that friends can behave badly no matter how impeccable their past behaviour to us has been.

Now, Hume was making a logical point about the essential incompleteness of induction. Yet both common sense and science tell us that induction is a practical matter and, as we will see in later chapters, the reason that ordinary physical experience seems to contradict Humean uncertainty lies in the scale of human life: Hume was correct both logically and factually, but to see that we have to look at events on a time scale very much greater than the one we are used to. Over a long enough period the seeming certainties of the physical world become as uncertain as the social relationships that ordinary experience tells us are infested with Humean uncertainty. That your house will one day cease to be where it is at present, if one adjusts the time scale, is as certain as that a friend or ally will betray you. As we will see, Hume's critique of induction is the beginning of our understanding of the evolution of intelligence.

This brings us back to the question as to whether philosophy or science is the appropriate vehicle for understanding knowledge. Science, as already stressed, is just a particular way of knowing, which proceeds by guessing at the nature of the world (theorizing) and then disciplining and revising those guesses by testing how they fit with the experienced world (observation and experimentation); in a sense, science combines rationalist and empiricist philosophies into an inseparable method. That being the case, it is hardly surprising that the answer to the question should be that we need both philosophy and science — though philosophers, whose criteria are of analytic excellence of argument rather than the pragmatism of science, may resent the incursions of science more than a little. They may even deny that science has consequences for our understanding of knowledge, but that is not a tenable position. What philosophy has done is refine the problems of knowledge without giving us answers that square either with ordinary life or with the extraordinary success of science itself, which (and the point bears repeating many times) is a form of knowledge. It is to get at those answers that we have to go to science itself. Science, after all, year by year, answers more and more questions about the nature of the world that we live in. If through science knowledge of the universe now extends back to within billionths of a second of its beginnings, and if through science the molecular basis of life is becoming known, then why should science not be equally successful in telling us about knowledge itself? The answer is that it certainly is successful in this regard now, if only a little, and will be even more successful in the future.

But which science? Well, it must obviously be some form of biology. And while divisions in science are most often arbitrary and sometimes quite meaningless, most biologists would agree that there is a distinction to be drawn between those areas of biology that deal with proximate causes and are pursued almost wholly as laboratory enterprises, such as physiology and bio-chemistry; and those areas of biology that are concerned to establish the ways in which living things change in time, and which roughly correspond to evolutionary theory and the empirical areas that feed such theory. The science of knowledge falls into both major divisions. On the one hand, psychology and the neurosciences deal with the proximate causes of knowledge. They tell us about the laws and mechanisms that govern individual perception, learning and memory, the embodiment of individual knowledge. Towards the end of this book I will briefly consider certain aspects of these. On the other hand, evolutionary theory itself is, as most of this book will argue, a theory of knowledge as well as a theory of transformation in time. It is a theory of knowledge because the process of evolution is the principal provider of the organization of living things, and one of the most important kinds of organizations of living things, adaptations, is related to particular aspects of environmental order.

At the start of this chapter I defined knowledge as a relationship between the organization of the brain and specific features of order in the world outside. Now I want to broaden that definition and argue that knowledge is the relationship between the organization of *any* part of a living creature's body and particular aspects of order in the world outside of that creature. What we normally in our everyday experience think of as knowledge, and what the philosophers mean by knowledge, are merely special (if particularly interesting) instances of this wider set of possible relationships. In so far as adaptations of all sorts are forms of knowledge, then evolution itself is the process by which knowledge is achieved.

It will become clear as we go through the book that by evolution I mean more than the processes that lead to new species in the sense that Darwin and others originally used the term. Indeed, we will see that Darwin himself realized that evolution is a process responsible for more than just the origin of species. It will be argued that evolution as a much wider process leads, among other things, to the thoughts and ideas that we have in our heads, and in that sense is directly responsible for our knowledge as commonly understood. It is also responsible for the contents of our cultures, such as our shared beliefs of what is good to eat, what we should read and how we should dress. However, it is to the theory of evolution in its more traditional form that we turn first.

SUGGESTED READING

Chisholm, Roderick M. (1989) *Theory of Knowledge*, 3rd edn, Englewood Cliffs, NJ: Prentice-Hall. A brief yet uncompromising and comprehensive tour through the philosopher's world of knowledge.

Cornman, James W., Lehrer, K. and Pappas, George S. (1987) *Philosophical Problems and Arguments: An Introduction*, 3rd edn, Indianapolis: Hacket. A general primer on philosophy with a good chapter on the problem of knowledge and the challenge of scepticism.

Russell, Bertrand (1961) *History of Western Philosophy*, 2nd edn, London: Unwin. A wonderfully erudite review of 2,500 years of philosophical thought.

2

Evolutionary Theory

There had been theories of evolution before 1859, which is the year in which Charles Darwin's *Origin of Species* was published. Only one of these, that of Jean Baptiste de Lamarck, is of any real interest and importance. So I will begin with Lamarck before going on briefly to consider Darwin's revolutionary ideas, as well as the near-death of Darwinism in the first part of this century and its revivification by what has become known as the modern synthesis. Since 1960, some disquiet with neo-Darwinism has appeared along with certain new ideas about evolution, and these will also be described. Then, in the final section of this chapter, we will move to the specific evolutionary notion that is of most importance to the understanding of the evolution of knowledge, that of adaptation.

Now, evolutionary biologists are as combative and argumentative a group of people as one is likely to find in science. So, lest any of my readers be evolutionists of one kind or another by profession, I had better declare myself at the outset. What follows is as close to neo-Darwinian orthodoxy as one can get in almost every way. The only serious deviation is to insist on the existence of evolutionary processes acting within individual creatures, and not just exerting a force on and between them. However, touched as I am by this heresy, I am not so unsympathetic to other, usually more heterodox views as to ignore or rule them out altogether as alternative approaches. In general, though, it will be apparent in what follows that I believe neo-Darwinism, together with its extension to what occurs in our and certain other animals' heads, to be the way and the light for the understanding of all biological sciences – and that includes the human sciences.

WHAT IS MEANT BY EVOLUTION?

I have often found myself embroiled in arguments with my colleagues after giving a talk, and frequently on such occasions sections of the audience have been at odds with one another as well as with me. At some point I will stop proceedings and ask people to say what they mean by evolution, and when they do the source of the disagreement is immediately obvious. People mean different things by the word. In order to avoid this kind of confusion here, we must sort out some of the most common and important distinctions that can be read into the word 'evolution' and make clear the meaning that will be used throughout the rest of this book.

The philosopher C.G. Hempel once suggested that one should distinguish between the story of evolution and the processes that drive evolutionary change. What follows in the next few lines is an example of an evolutionary story.

Early reptiles gave rise about 270 million years ago to a number of descendent lineages. One of these led to modern snakes and lizards; another to a varied group of animals that included the dinosaurs and, some believe, modern birds; yet another were a small and rather insignificant group of creatures known as therapsids. When the dinosaurs became extinct about 65 million years ago, perhaps for the same reason that many other animals and plants became extinct at that time, a small, active, hairy group of animals, descendants of predatory therapsids, that had existed alongside the dinosaurs, suddenly rid of the fearsome competition of the latter that had forced them to live in obscurity for many millions of years, were able to expand into the ecological niches vacated by the then extinct dinosaurs and thus to become one of the major forms of life on Earth, the mammals.

End of this particular evolutionary story. It is clear that what Hempel meant by an evolutionary story is a descriptive account of the history of life which says little or nothing about the causes of these events.

Now, few scientists would confuse descriptive accounts with causal accounts. Yet many do confuse the consequences of evolution with its causes. For instance, a definition of evolution commonly used by biologists is that of descent with modification, meaning that from one generation to the next creatures may come to differ from their predecessors. This is similar to the Darwinian claim that evolution is the answer to that 'mystery of mysteries', the origins of species. More technical is the notion that evolution is properly understood as the change over time in gene frequencies of a breeding population (I will say a little below as to what this means and why it is important). Then again, there is the view that evolution is the explanation of adaptations. All of these are correct, at least in one sense. However, each is a statement of the consequences of evolution. Descent with modification, speciation, changes in gene frequencies and adaptation are all consequences of evolution. They are not what evolution *is*, which is the process or processes that *cause* these various effects. For the rest of this book, the term 'evolution' will be used to mean the processes by which living things change in time. Evolution *is* a specific set of processes. Thus prepared, we can now consider what these processes are.

LAMARCK

Scientists are notorious for not changing their minds as they get older. One of the remarkable things about Lamarck (1744–1829) is that he came to accept the idea that species become transformed over time at the turn of the nineteenth century when already in middle age. He probably did not, however, pluck the idea from the air, since the non-fixity of species was widely discussed at the time. Darwin's own grandfather, for instance, had written a curious but popular book, published in England in 1794, which expounded in small part upon the notion of the transformation of species. And all the while evidence was gathering that was eventually to prove grist to the evolutionists' mills.

Perhaps the most important was the realization in the eighteenth century that the Earth is very old, to be measured in millions of years at least. Kant had suggested an indefinitely old planet in some of his writings, and tens of millions of years on other occasions. Darwin was later to estimate thousands of millions of years, which was remarkably accurate – we now know the Earth to be about 4.6 thousand million years old. The important point is that these quite drastic departures from the biblically inspired estimates of just a few thousand years were essential to the conceptual baggage of anyone toying with the idea of evolution because of the long time periods that evolutionary processes would require.

Also significant in the eighteenth century was the growing appreciation of the extraordinary diversity of living forms. Decade by decade the numbers of species known and estimated increased sharply, and the great classifiers like Linnaeus proceeded to construct hierarchies of relationships between animals and plants that strongly suggested common origins and subsequent diversifications. Another important realization of the time was that extinctions of species had occurred. Creatures like mastodons and mammoths, which had once lived, no longer do so. If living types had been lost, and lost perhaps continuously over very long periods of time, how could this be reconciled with the seeming increasing diversity of life? Surely if life forms had been rapidly created, as in the biblical account, and subsequently subjected to gradual extinction, the diversity of life on Earth should have decreased over time, not increased. Finally, around the turn of the nineteenth century the science of palaeontology (the study of fossils) was becoming established, and the general pattern that began quickly to emerge was that the older a fossil was the less it resembled contemporary forms, whereas more recent fossils were more similar to living species. How was this to be explained?

Perhaps the decisive experience in turning Lamarck into an evolutionist was his appointment to a professorship in 'inferior animals', or what we now refer to as invertebrates (animals without backbones, including worms, insects and snails). Invertebrate animals contribute much the greatest amount to the Earth's diversity of living forms and are present in very large numbers indeed. No group was more likely to tempt biologists away from the dominant ideas of the time regarding the origin and diversity of species. These dominating ideas were, on the one hand, the familiar one of divine creation and, on the other,

Platonic essentialism, which postulated discrete and discontinuous forms, both living and non-living. Roundworms are roundworms, and flatworms are flatworms: there are no intermediate forms and there is no continuity of type. This is essentialism as applied to biology. Ernst Mayr, doyen of living evolutionists, argues that essentialism was the conceptual strait-jacket that kept virtually all eighteenth-century scientists away from the revelation of evolutionary understanding. And those who were not paralysed by essentialism and able to believe in the transformation of species seemed quite unable to offer ideas about how species might be altered.

Well, whatever the decisive experience might have been for Lamarck, he became a believer in evolution and was the first to offer a theory of how the transformation of species occurs. His theory can be most easily stated in a series of points.

(1) All organisms are wonderfully well adapted at all times and in all places. This, of course, was not an observation unique to Lamarck. It was a commonplace, and we now know erroneous, belief held by all natural theologians of the eighteenth century, and before.

(2) All environments change in time. This relatively novel idea originated in eighteenth-century science, and was later to form the basis of an important nineteenth-century debate, primarily among geologists, about the patterns, causes and rates of change of conditions on Earth.

(3) It follows from the first two points that organisms must change in time.

(4) The way this occurs is by the organisms responding to environmental change which imposes new needs with new activities that result in modifications of body structure. This chain of events is sometimes referred to as the law of use and disuse, because, it was believed by Lamarck, the less used body parts would be reduced and wither, whereas the more used and exercised structures would grow and transform.

(5) Such changes in structure are passed on to the offspring. This is usually referred to as the principle of the inheritance of acquired characters, and is the feature of the theory that has come to be

most closely identified as Lamarckian. In fact, it is the least original part of the theory, being a piece of folklore, subscribed to, in one form or another, for millennia.

(6) Governing all these mechanisms is the overriding cause of evolutionary events: the tendency, immanent in all living things and divinely inspired, for change from simplicity and imperfection to complexity and perfection of form. This conception is rooted in the ancient notion of a *scala naturae*, where life begins spontaneously in slimes and moulds and then progresses, in Lamarck's theory by the mechanisms outlined in points (4) and (5), through stages of intermediate complexity such as those manifested by insects and fish, to its most complex and perfect form, human beings.

A theory less of descent than of ascent, Lamarck's relatively complex set of ideas was, none the less, an attempt to provide an account of adaptive fit between organisms and their environment, points (1)–(5), and of species transformation, point (6), dressed up in processes and mechanisms. And as we will see, Lamarck's ideas contained a core notion that is effectively one of only two ways by which evolution can occur.

For these reasons Lamarck played a key role in the evolution of evolutionary theory, and it is one of the oddities of science that Lamarckianism has become a term of abuse, for though we now know that Lamarck was wrong in every way, being wrong in science is not usually a reason for dishonour. It is in the nature of science that most scientists hold erroneous views a good deal of the time, and none are correct all of the time. In any event, Lamarck did eventually gain a following, and Lamarckian theory had some prominence in science, especially the social sciences, until the birth of the modern synthesis in the 1930s. As a coherent theory, however, it was always weakened by the fact that its principal source of supportive evidence, the supposed harmony and perfection of living forms, was the same as that claimed by the natural theologians like William Paley for proving the existence of a wise and omnipotent Creator. A theory of evolution which draws support from the same evidence as theological claims cannot have much of a future. It is remarkable that Lamarck's theory endured as long as it did.

DARWIN

Darwin's theory of evolution was vastly superior to that of Lamarck, partly because Darwin detailed so much evidence to support the idea that species change. In Darwin's hands, evolution became science rather than fanciful speculation. And this was partly because one of the great strengths of Darwin's theory, and hence one of the most important differences between it and Lamarck's theory, was that it considered imperfection to be central to evolution. To be more exact, Darwin recognized that living things, even of the same species or variety, exist in a great diversity of types which, contrary to Platonic essentialism, grade into one another, often imperceptibly. Apart from identical twins, members of the same species do not all share identical characteristics, and there are no discrete boundaries demarcating one species from some other related species. Further, some individuals are better fitted for survival and reproduction than are others, and those less fit are eliminated. This was Darwin's principle of natural selection, one of the great scientific insights of all time.

In a famous passage in his autobiography, Darwin described how he made the transition from his knowledge of artificial selection, as practised by breeders of horses, dogs and other domesticated animals, to an understanding of how species change in time:

I soon perceived that Selection was the key-stone of man's success in making useful races of animals and plants. But how selection could be applied to organisms living in a state of nature remained for some time a mystery to me. In October 1838, that is fifteen months after I had begun my systematic enquiry, I happened to read for my amusement 'Malthus on Population', and being well prepared to appreciate the struggle for existence which everywhere goes on from long-continued observation of the habits of animals and plants, it at once struck me that under these circumstances favourable variations would tend to be preserved and unfavourable ones to be destroyed. The result of this would be the formation of new species.

Here then I had at last got a theory by which to work; but I was so anxious to avoid prejudice, that I determined not for some time to write even the briefest sketch of it.

In the most extraordinary case of scientific caution known, Darwin did not publish his theory until 1859 – and he published it then only because he was forced into it by the fact that A.R. Wallace, another English naturalist, had independently discovered the principle of natural selection. The essence of Darwin's theory revolves, according to Ernst Mayr, around a small series of facts and inferences. (I am entirely indebted to Mayr's several marvellous analyses of Darwin's thinking for the next few paragraphs.)

Available to Darwin were five 'facts'. The first was the potential of any population of animals or plants for very rapid increase in size, technically referred to as exponential growth. For example, each female Atlantic cod produces millions of eggs each year. If even only a small fraction of these were to mature into adult numbers and themselves each produce millions of offspring each year, the oceans of the world would not be large enough to contain the numbers that would result in just a few short years. The second 'fact' – in the nineteenth century more a belief really – was that despite this potential for massive growth, most populations maintain a relatively constant size. Third, the resources available to any population are always limited. These facts made sense to Darwin in the light of the Malthusian inference of a struggle for existence between individuals, and they gave him a very particular vantage-point from which to muse and ponder upon the significance of two additional pieces of information available to him.

The fourth fact was the nineteenth-century realization of the uniqueness of individuals. All elephants may look alike to the casual observer, but careful measurement or the practised eye reveals that each is unique and unlike any other. This, of course, is a commonplace observation about human beings, but it is true of elephants as well – and indeed true also of snails and palm trees.

The final fact, already referred to above as well as in the quotation from Darwin's autobiography, was that the individual differences that make parents unique can be passed on to offspring. This heritability of variants, as it is technically referred to, was well known to animal and plant breeders who practised rigorous artificial selection as a way of controlling inheritance to produce in the offspring such desirable attributes as speed of foot, coat colour or fierceness of temperament.

Darwin's genius lay in his ability to combine these facts with Malthus's inference to arrive at two further, and original, inferences. The first of these is known as the principle of natural selection. Darwin reasoned that any living creature is a patchwork of variable traits which together somehow determine how 'fit' that creature is for survival and reproduction in a particular environment. In elephants, for example, there will be variation in the lengths of tusk and trunk, acuteness of smell and hearing, susceptibility to certain diseases and loyalty to other elephants. The mix of traits will determine for how long a creature will survive and how many offspring it will produce. So, no matter how magnificent and strong are the tusks and trunk of an elephant, it will not be a fit animal if it readily succumbs to a common illness. And elephants that are resistant to such illnesses, and also possess sufficiently strong tusks and trunks, are more likely to survive to an age at which they can reproduce. The offspring will carry at least some of the characteristics that rendered their parents capable of survival and reproduction.

Those creatures that are not fit (for whatever reason) will not survive and reproduce, or produce only a small number of offspring, and therefore will not propagate their characteristics down through the generations. In the case of our elephants, the offspring of fit parents will inherit the characteristics that proved so beneficial, and over time the population of elephants will come to be dominated by animals that combine such advantageous characteristics.

So that we are quite clear about what this means, allow me to thrash this out in a bit more detail. Let us say that a normally harmless micro-organism present in the environment of the elephants undergoes a change to become the cause of fatal or debilitating illness. Quite independently of the micro-organism and the changes that it might undergo, a small number of elephants develop a slight change in physiology that allows them to cope with the new deadly disease. This change in the elephant is *independent* of events in the elephants' world. It is *not* a response to the changes in the micro-organism. Such a change might have been appearing spontaneously every now and then over thousands of years but not lasting in the population because it served no function. With the change in the micro-organism, a small number of elephants now have a constitution that enables them to survive the illness. If the resistance to the disease is inherited by the offspring, then over time natural selection will lead to a change in the elephant population in at least this one important respect: where previously only a small proportion of the elephant population had resistance to the disease, in time, because of differences in survival and reproduction, and because the characteristic responsible for disease resistance is inherited, almost all of the elephants will have this particular trait. Natural selection will have wrought a change upon this population of elephants.

Darwin's second inference was that over long periods of time natural selection may cause populations to change sufficiently so as to become new species. For example, and as I have already said, elephants differ from one another in many different ways. Not only does their susceptibility to disease change over time, but so too might any other characteristic be shaped by natural selection, including dietary preferences, length of hair and love of little elephants. As the world changes, which is the absolutely indispensable and fundamental condition for evolution to occur, so too are changed the demands that are made on the living things that are able to survive and reproduce. Change is not, however, always universal or identical in kind everywhere. It may, and does, happen that change of one sort or another is localized to a particular region. In that event, local populations will diverge from one another. Eventually, separated populations of elephants will be so different from one another that if they meet, members of one population will be unable to mate with members of the other or, if they do, the offspring will not be viable. The result is two species of elephant where before there was only one. (The story of the evolution of the African and Indian elephants is almost certainly of this kind.) In time, and with further change, one or both species may fail to adapt, and in consequence may become extinct. Putting all this in modern terms, small, gradual and adaptive changes occur in isolated populations (referred to as microevolutionary changes), and these may eventually result in speciation (macroevolution).

Darwin's theory was thus able to account for the relatedness of all living creatures, their distribution in space, for the characteristics of the fossil record and for extinctions, as well as for the origins of new species and the existence of adaptations (more of which below). It was, and remains, an astonishingly powerful theory and, of course, the most controversial and bitterly opposed scientific theory of all time.

DARWINIAN VS LAMARCKIAN EVOLUTION

However wrong we may now believe Lamarck to have been, his theory does predict transformation in time, and in that sense, it is a plausible evolutionary theory. So Lamarckianism must be taken seriously and contrasted with Darwinism. Indeed, at an abstract level of description, Darwinian theory and Lamarckian theory offer the only two forms of evolution that have ever been presented. All else derives from these fundamental types. Darwin's was a selectionist theory of evolution, whereas Lamarck's was instructionist. The essential difference between them is as follows: for Darwin the traits that contribute towards the overall fitness of an organism occur initially by chance, prior to their being 'required'. They are then selected from the multiplicity of variants and transmitted to offspring. In other words, the process by which adaptive traits are produced is initially independent of their potential adaptive usefulness. And it is in this sense that Darwinian evolution is often referred to as blind or undirected. We will see shortly that even a modern, and some think radical, variant of evolutionary theory, entitled punctuated equilibrium theory, conforms in this way to Darwinism. Compare this with Lamarck, for whom adaptive traits are produced after and in response to the environmental changes for which they are required. For this reason Lamarckian evolution is thought of as directed, with the production of adaptive traits being instructed by environmental events.

In terms of mechanisms, what Lamarckian evolution requires is a highly malleable substrate that can be modified by the environment and then transmitted to offspring – hence the inheritance of acquired, instructed characteristics. Darwinian evolution is driven by mechanisms that generate great diversity of characteristics, by some means of selecting between them and then by a device for propagating the selected variants to offspring.

These differences are very important, and we shall return to them in the chapters on learning and intelligence. For the rest of this chapter Lamarckianism will not be mentioned again, because the processes and mechanisms that lead to speciation and the adaptive form of individual organisms are Darwinian. Occasional flurries of excitement and activity in the scientific literature notwithstanding, there is no good evidence of any kind from any source that Lamarckian processes operate in evolution as traditionally understood.

THE NEAR-DEATH OF DARWINISM, THE RISE OF GENETICS AND THE MODERN SYNTHESIS

Quite apart from sectarian religious opposition to evolutionary theory, some of which is still with us, Darwin's ideas did not win immediate universal acclaim among scientists. Even his staunch friend and ally, T.H. Huxley, expressed unease at certain features of Darwin's theory, especially its insistence on evolution being a slow and gradual process. Darwin himself was troubled by two related aspects of the theory. He did not know the source of the variation that the theory demands, and he had no idea as to how selected variants are passed from parents to offspring. The notion of blending, at that time the principal idea about inheritance, held that offspring combined the attributes of parents to give rise to a kind of average of parental characteristics. This was exactly what Darwin did not want, since it predicted the elimination of variation, not its constant generation.

Yet around the time that *The Origin of Species* was published, experiments were being conducted in Moravia by a little-known monk named Gregor Mendel that would begin to solve Darwin's problems. Mendel worked very much on the fringe of science. He published little and lectured on his experiments on just a few occasions, to small and insignificant gatherings. His main work was conducted over nearly a decade beginning in 1856, and he ceased all scientific work in 1871. Darwin never knew of his work, and it is doubtful that Mendel himself ever really understood its significance.

In brief, what Mendel discovered, by a mixture of good luck in his choice of an experimental subject (the edible pea) and considerable scientific skill, was that the characteristics of organisms are controlled by single factors, as he called them, contributed in sexually reproducing organisms by both parents; that these factors combine together in ways that give rise to the characteristics of offspring; and that these factors do not blend but retain their integrity after fertilization and are passed on to offspring, even if their effects are not manifested in the traits of the individual. We now call these factors genes, and the variant forms that a gene may take are known as alleles. So, in the classic example of human eye colour, there is a gene for this characteristic that may come in an allele giving rise to brown eyes and another allele that leads to blue eyes. The offspring of parents one of whom has brown eyes and the other blue eyes has an eye colour that is not some muddy blend of brown and blue but either brown or blue. And if the offspring has brown eyes, that does not mean that the allele for blue eyes has been lost or corrupted; it is retained intact and may be passed on to the next generation of offspring, some number of which may express that allele in the form of blue eyes.

Furthermore, Mendel's laws revealed the cause of individual variation seen within populations, even if Mendel himself little understood the nature of the underlying mechanisms. For example, we now know that Mendel's factors, the genes, are chemical structures contained on the chromosomes of cells and that the chromosomes are present in creatures like ourselves in pairs. *Homo sapiens* has twenty-three pairs of chromosomes. When sex cells – eggs or sperm – are formed, the pairs of chromosomes are divided out into separate cells so that each sex cell receives only twenty-three single chromosomes. This reducing division – known as meiosis – is essential to prevent the number of chromosomes from being doubled at each fertilization. Now, during meiosis, which member of each pair of chromosomes goes into a particular sex cell is independent of which individual chromosomes of all the other twenty-two pairs go into that cell. (In fact, the segregation that occurs is not quite so straightforward: in the process known as cross-over, the members of each chromosome pair exchange parts with one another, which actually increases the variability that occurs.) Such independent segregation alone gives rise to 2^{23} possible different sex cells – which is a very, very large number. There are other genetic devices that add to the production of variation which is so prodigious that there is now good reason to believe that, with the exception of identical twins, no two humans of the same genotype (genotype is the term given to the totality of genetic structure possessed by an individual) have ever existed in the history of our species. Thus Mendel, and the subsequent development of genetics, revealed the source of variation within a species and how those variations are inherited.

Mendel's work was discovered independently at the turn of the century by several scientists and formed the basis of the new science of genetics. The existence of chromosomes, the arrays of genes on chromosomes, the ways in which the structure and expression of genes are changed by mutation, the linkage of genes, the characteristics of cell division and the cross-over of chromosomes, the manner of gene expression in the individual and how such expression is affected by the presence of other genes – all these and more were discovered by geneticists during the early decades of this century. Genetics rapidly grew to be one of the most successful sciences of that time and generated such excitement that Darwin's principle of natural selection as one of the main driving forces of evolution languished in a state of neglect. Indeed, some openly dismissed natural selection as not having much of a role at all in evolution. Genetics alone, it was thought, could provide the proper understanding of speciation and the origins of adaptation. Darwin's gradualism was momentarily eclipsed by mutationism and saltationism. After all, the geneticists were getting to the point where they could actually specify the mechanisms by which nature might make jumps from one form to another.

Then, in the late 1920s and early 1930s, Darwinism was revived by the work of several theorists – principally R.A. Fisher, J.B.S. Haldane and Sewell Wright – who achieved, in what became known as the modern synthesis, a marriage between genetics and the Darwinian theory of evolution. The initial emphasis of the synthesis was on mathematical descriptions of the way gene frequencies within a population would change, the changes being *caused* by natural selection. If, for example, one allele of a gene is passed on to 100 offspring in a population, whereas another allele is passed on to only 99 offspring, then the first allele is said to have a selective advantage of 1 per cent relative to the second allele. In time, the allele with the selective advantage will increase in frequency in that population. In principle, selection coefficients can be attached to all genes, whether positive or negative, their magnitude depending upon the environment of the population; and hence the changes in gene frequencies of the population can be understood in terms of the selection pressures exerted upon it. Since the genetic constitution of a population is expressed in the form of individuals making up that population, referred to as phenotypes, changes in gene frequency mean a change in phenotypes. Hence change in gene frequencies becomes an important measure of the microevolutionary changes in that population; and it is microevolutionary change that drives macroevolution.

Thus the modern theory of evolution, sometimes referred to as neo-Darwinism because it brings together the notions of selection and genetical change, sees evolution as a two-step process. The first step is the generation of variant phenotypes, at least in part because the genetical machinery ensures much variation. The second is the selection of phenotypes and their differential reproduction. These microevolutionary events, if confined within a breeding population, will lead to changes in the form of that population that might eventually lead to macroevolution, that is, the formation of a new species.

Subsequent developments in the modern synthesis focused upon the details of types of selection and the genetic structures of populations that lead to speciation; patterns of speciation; the origins of genetic variability; adaptations and large-scale adaptive patterns that make up what are known as life-history strategies; and, with the coming of the age of molecular biology in the 1950s, increasingly detailed descriptions of variation at the molecular level in natural populations. By 1959, a century after the publication of *The Origin of Species*, the theory of evolution, essentially Darwinian in nature, was firmly established. It was thought to have great explanatory power and to be the central theorem of the biological sciences. 'Nothing in biology makes sense except in the light of evolution,' wrote Theodosius Dobzhansky, one of the architects of the synthesis.

RECENT DEVELOPMENTS

In 1982, when a number of international conferences were held to mark the centenary of Darwin's death, the sense of certainty and self-satisfaction of 1959 had been replaced by a more questioning attitude as to the adequacy of the synthesis. I want briefly to consider here the reasons for this change. Now it is never easy in science to sort out the truly significant from the trivial in present-day theory, especially when one well knows of instances in the past when what at first sight seemed to be radical departures were subsequently incorporated into mainstream theory as a variation on a theme. However, such sorting out of current theory is what I must do, and in order to keep the discussion as short as possible, but without losing the flavour of the intensity and vigour of the debate, I am going to have to lump certain views together, without, I hope, doing too much violence to their arguments.

It must be said at the outset that no biologist of any stature whatsoever denies in general the place of evolutionary theory as the central explanation of the form and origins of living things. There have, it is true, been some attempts to recast evolutionary theory in very radical ways. Some of these have centred about a set of ideas relating to the organization, co-ordination, self-regulation and transformation of complex structures – a movement that flies the flag of structuralism or constructivism. Others are associated with non-equilibrium thermodynamic theory and are concerned with analysis of the transformation of structures that are far from thermodynamic equilibrium and which maintain their orderly structure by exploiting energy resources external to themselves. Having flagged them in this way for the interested reader to pursue in other places, I will not deal with either of these schools any further beyond a statement as to why I am dismissing them so cavalierly. The structuralists, in my view, are merely waving their hands in suitably complex fashion but not saying anything that has either lucidity or real theoretical clout. I am more likely to rue my stance on the non-equilibrium thermodynamic approach. It seems to me that they are rewriting evolutionary theory in physico-chemical terms, which may or may not be an interesting exercise, but which certainly loses the biological flavour of the theory. Evolution is not a theory of physics. It is a – nay, *the* – central theorem of biology. As such it should be written in the language of biology and not, by translation, in the language of physics and chemistry. But however much structuralists or non-equilibrium thermodynamic theorists might disagree with these statements, none would argue that evolutionary theory is trivial or plain wrong. The real dispute concerns content and style.

So it is that the substantial arguments have centred not on any claim that neo-Darwinism is wrong or that it needs to be written in a different language, but on the claim that it is incomplete. For convenience it is possible to divide the disputants, apart from the traditional neo-Darwinists who are inclined to believe that their views cannot be improved upon, into two camps. One of these comprises people who think that certain essential mechanisms and concepts must be added to Darwinian selection. The other insists on the expansion of evolutionary theory, usually into one taking some kind of hierarchical form. These are not mutually exclusive positions.

Let us start with those who believe that evolution comes about through more than just natural selection acting on phenotypes to sort and sift between genetic alternatives, perpetuating the selected variants from one generation to the next. As might be imagined, several mechanisms in addition to those of neo-Darwinism have been suggested and, again for simplicity, these can be broadly grouped into two. The first are concerned with modern molecular biology, and the second with the general notion of constraint.

The rate of growth of knowledge in molecular biology that has occurred since 1953 (when the molecular structure of the genetic material, DNA, was first described) has been quite extraordinary, and probably without parallel in the history of science. The conception of genes as discrete entities neatly arranged along the lengths of chromosomes – sometimes referred to as the 'beads-on-a-string' idea of genetics – has had to be relinquished. Genes are chemical structures of great complexity which are smeared across the chromosomes rather than being neatly and linearly packaged. They can only be properly understood if seen as entities with significant internal structures and grouped in complicated ways into sometimes spatially widespread functional units called multi-gene families; and if it also be realized that genes are not simply passive providers of information that retain their structure across generations, but reactive complexes that are in constant and dynamic interaction with their carriers, the organism. That is, the structure and function of genes in body cells can be altered by these gene–phenotype interactions.

Viewed in the light of this kind of complexity, it becomes possible to entertain the idea of the existence of evolutionary causes beyond that of natural selection acting upon the consequences of genetically caused variation of the phenotype.

Since the general point is crucial to an argument about the causal role of intelligence in evolution developed in a later chapter, it is worth quoting a molecular biologist, John H. Campbell, at some length:

When Darwinism and even modern neo-Darwinism were formulated, the structures of organisms and genes were so vaguely known that evolutionists deliberately denied their roles to avoid the specter of vitalism. Instead, evolutionists reverted to the mechanical paradigm of physics, in which inert objects move only in passive response to exogenous forces pushing upon them from the outside. Evolution became change that the external environment forces upon the hapless species instead of a function that organisms are structured to carry out. In this Darwinian perspective, species do not evolve in an active sense; they only get evolved by an external natural selector. They do not have causal roles in their evolution.

During the past decade, molecular biology has advanced genetics to a structure-function science. This progress is forcing us to recast evolution as well . . . The profound new discoveries for evolution no longer concern the demands of the environment on species, but rather the molecular structure of organisms themselves. These discoveries raise profound questions as to how biological structures function to direct, cause, and carry out evolution.

This massive increase in understanding of molecular genetics emanating from a knowledge of genetic structure has led to suggestions that go beyond the conception of the genes as just conveyors of information between generations and as generators of variation, crucial as these functions might be, and that go beyond the effects of the environment and natural selection on this variation. An early instance of such thinking appeared in the late 1960s. It was suggested that the genetic constitution of a species might be altered not only by selection but by a process of 'random walk' of genetic structures resulting from neutral alterations of genetic material. The term 'neutral' was used to indicate that such change might not necessarily be sorted and sifted, at least not immediately, by selection. That is, many of the alterations in genes are neither positive nor negative in selectional terms. Indeed, they may not, at least initially, be expressed in phenotypic structure. Thus sheltered from the scrutiny of selection, quite marked changes in gene pool constitution might occur that are not attributable to the effects of selection. Most evolutionists now, even dogmatic neo-Darwinists, accept that a 'neutral' form of genetic change does occur. The argument about its extent and significance remains.

In the early 1980s, this position was taken further in what is now referred to as molecular or evolutionary drive theory. The argument goes like this. Given the complexity of molecular genetic mechanisms, at least some alteration in gene structure comes about not just through the action of selection on expressed genes or through the chance occurrence of neutral mutations, but as a result of the structural complexities of the genome, i.e. the entire genetic complement of each individual, itself. That is, as expressed in the John Campbell quotation above, there are forces that operate as an intrinsic part of genetic structure that are a source of autogenous change in genetic form. For example, it has long been known that not all genes assort independently during the reducing division that occurs in the formation of sex cells. Some genes are linked, occurring on the same chromosome. Such linked genes may interact with one another when chromosomes break and rejoin, a process known as cross-over. Cross-over may result in the frequency of one gene or allele increasing at the expense of another when the DNA strands become physically intertwined with one another. This phenomenon, known as gene conversion, will lead to changes in gene frequency without regard to the selection pressures that are operating on the organisms in which gene conversion is occurring.

This is currently a contentious issue (as are *all* theoretical issues that are now being advanced), but my money is on the existence of these internal genetic forces. The complexity of molecular genetic mechanisms is such that it seems unlikely that such effects would not exist.

Now, it must be said loud and clear that neither the 'neutral' nor the more active 'driver' theories of genetic change are anti-Darwinian. They do not deny the existence and importance of selection in directing genetic and evolutionary change. What they do is add another source of change to supplement the variation–selection formulation that is so central to neo-Darwinism. And this brings us to the issue of constraint, which in a sense is opposite in effect to that of neutralism or evolutionary drive, because the argument is that there are limitations on the variation that evolutionary processes, including selection, have to work on at any moment in time. Neo-Darwinism is sometimes caricatured as a theory of blind chance that can result in anything: given the appropriate selection pressures a mouse could evolve into a man. This is a half-truth because what is left out of the statement is the improbability of its occurrence. The evolution of all living forms has been improbable, of course, and while it is important to recognize the chance-laden nature of the evolutionary process, we don't normally comment on evolutionary events in a 'what might have been' manner. The chances of a mouse evolving into a man are vanishingly small because it would require not only the right sequence of selection pressures, but also the right array of genetical variants and, just as importantly, the right conditions for driving viable and functional structural change. That really makes the mouse-to-man caricature much less than a half-truth and closer to a plain false depiction – though neo-Darwinists are not entirely free of blame in the way their theory has been misrepresented, because they have been so insistent on the central role of selection, and more than a little reticent in accepting that other forces might be operating to produce evolution.

Be that as it may, not only is the genome structurally complex, but so too are the phenotypes, the actual living things, into which the genetic structures develop. Such complexity inevitably constrains the changes that can be borne by a complex system if that system is to retain its functional integrity. This is a twentieth-century way of referring to what in the nineteenth century was called the 'unity of type', which Darwin, in *The Origin of Species*, defined as 'that fundamental agreement in structure which we see in organic beings of the same class, and which is quite independent of their habits of life'. It is significant that Darwin added that 'unity of type is explained by unity of descent'; for not only does complexity impose constraints and hence account for the tendency of living creatures to conform to some limited number of basic structural plans (*Bauplan* is the word used by nineteenth-century

German natural philosophy), but evolution always proceeds in time, and history acts itself to constrain the variations that can be generated. Initially a mouse can only generate variants that are mouse-like, because a mouse is a prisoner of its own history, and also because of the structural constraints of its genome and the developmental constraints by which a mouse phenotype develops from a mouse genotype. Constraints, in short, are an expression of the limited tolerance that complex structures have of deviation, and beyond which they lose their functional integrity. Because constraints are a universal characteristic of any functional, complex structure, the selection of constrained variation is a more accurate description of what is at the heart of the evolutionary process.

And what of proposals to expand evolutionary theory such that it takes into account the idea of hierarchy? In 1972, the American palaeontologists N. Eldredge and S.J. Gould proposed what they called the theory of punctuated equilibrium, which in several ways does offer a serious departure from neo-Darwinism. One of their arguments was that a central claim of Darwinism, namely the gradual but continual pattern of evolutionary change, is wrong. What characterizes the fossil record in most cases, they said, is a pattern of relatively little change in a species (stasis or equilibrium) over perhaps millions, even tens of millions, of years, with change occurring relatively rapidly over periods to be measured in tens of thousands of years. In Gould's own words, the theory 'focuses upon the stability of structure, the difficulty of its transformation, and the idea of change as a rapid transition between stable states'. Now, this is not an easy argument to resolve, because neither Darwin nor his followers ever suggested that evolution occurs at some precise and unvarying rate. Quite the contrary, neo-Darwinism has good ways of explaining the 'jerks', the deviations away from smooth rates of evolutionary change. And for their part, the punctuationists, though saying that speciation involves rapid change, mean 'rapid' only in geological time. You have to be a palaeontologist to think that 50,000 years is a short time.

So despite comments such as that of Gould above, the proof or disproof of punctuation theory will not lie in the patterns of the fossil record. What will decide the matter are other features of the theory, of which two in particular are important. First, whatever the pattern of evolution according to the fossil record, what do punctuationists say *causes* the process of evolution? Usually the answers are rather negative, especially the denial that the theory makes any claim about how evolution comes about. Another negative claim, and one of considerable importance, is that the theory does not support the linkage between microevolution and macroevolution. That microevolutionary change occurs and is driven by selection is not denied. What is claimed is that the causes of speciation are confined to the macroevolutionary – that is, species – level.

This is where punctuated equilibrium theory makes contact with the notion of hierarchy. A hierarchy is an ordering of entities, at least a part of the ordering being dependent upon scale, that is, some dimension such as size, energy level or frequency. Now, a source of great confusion is the often unrecognized existence of two seemingly quite different kinds of hierarchy. The one, called a structural hierarchy, is characterized by a feature known as containment, or the Chinese-puzzle characteristic. That is, you take something, open it up and look inside. There you find another entity, which you open up and find that it, in turn, contains another entity. You keep doing this until you reach an entity that you cannot open up: at that point you are at the fundamental level of that particular structural hierarchy. The most obvious example is ourselves. Open us up and you find organs, inside of which are tissues, which contain cells, in which are to be found organelles, which are made up of macro-molecules. At the other end, each person is part of a social grouping or community, and we live together with other organisms to form a larger ecological unit. All organisms, not just humans, are structural hierarchies scaled by size and the strength of the binding forces that cause each level in the hierarchy to cohere.

It might be noted, incidentally, that classificatory hierarchies, exemplified by the Linnaean classification scheme (*Homo sapiens* is a species in the genus *Homo*, which is a part of the family Hominidae, and so on), are a form of the structural hierarchy. The

Linnaean classification is, however, a special conceptual case of a structural hierarchy and need concern us no further.

The second kind of hierarchy, a control hierarchy, is much closer in meaning and form to the original conception of hierarchical structure, which described the relationships of the angels to the Lord. Consider a crude sketch of how an army works. The generals issue orders to their field commanders, who in turn command lesser officers, who in their turn tell lower-rank people what to do. Here the scaling is one of authority, and the characteristic of containment is absent. The generals are not in any sense made up of their officers, and the latter do not contain private soldiers. Another important feature of control hierarchies is that they are much more dynamic than structural hierarchies. After all, generals do not just shout their orders into the wind. The soldiers report to their officers, who in turn communicate with the generals. The privates may even, on occasion, pass information directly to the generals. As a result of these interactions, the behaviour of the generals is altered in some way such that their subsequent orders are different from what they would have been in the absence of any feedback from lower levels in the hierarchy. Common sense tells us that this must be so if the activities of the whole system are not to become uncoordinated and incoherent. Thus control hierarchies are characterized by complex causal interactions between levels, whereas for structural hierarchies the principal interactions occur within levels.

Control hierarchies are a commonplace of human social systems, as well as of many other social species of vertebrate animal. They are also common in the brain. For example, we analyse visual information using a control hierarchy that extends from the retina, through a number of subcortical nuclei, and then on to large areas of cerebral cortex. Complex interactions are as characteristic of the control hierarchy of vision, within the visual system, as they are of the hierarchy that exists within a government ministry or a factory. The development of multicellular organisms from the genetic instructions of the sex cells also involves a cascade of control hierarchies.

The relationship between structural and control hierarchies is an extremely important theoretical issue. It is one of the central conceptual puzzles that have to be solved if the biological, cognitive and social sciences are to be married within the kind of grand synthesis that is occurring in the physical sciences. If science is a matter of carving nature at its joints, then the business of the biological and social sciences is that of carving nature at its hierarchical joints. The concept of hierarchy is central to everything in this book and we will return to it many times.

Neo-Darwinism has never been built upon a foundation of hierarchy theory. It is a two-stage formulation with genetics as the principal engine of variation and selection acting mainly upon the phenotype. But a stage is not a 'level', and sequence does not constitute a hierarchical formulation. No scaling relationship between genes and phenotypes is implied, and the whole way that these 'levels' have been presented has never been in the spirit of hierarchy theory. Over the last few decades, though, there have been a number of specific, and largely unheeded, calls for evolutionary theory to take the idea of hierarchy seriously. The general argument has been that because neo-Darwinism is a one-level theory, it is incapable of accounting for development, ecology and cognition within a single explanatory framework. Some of the best of neo-Darwinists have retorted that this is not evolutionary theory's role; that neo-Darwinism quite properly has a much more limited explanatory horizon. I disagree. I think that all of science aims to explain an ever-larger set of phenomena with an ever-smaller number of principles, and that the way towards the expansion of evolutionary theory is via the concept of hierarchy. Some punctuationists have begun to follow that route.

Punctuated equilibrium theorists do not deny the importance of genetic mechanisms, selection and adaptation as a *part* of the evolutionary process. They do, however, believe that speciation and the maintenance of species once formed are a problem in evolutionary processes acting at a higher level than individual organisms. Species selection is what is occurring. That is, species themselves somehow appear in variant forms and only some survive for any significant period. Species are a part of a structural hierarchy in which individual organisms are one of the lower-order components. The sorting of successful from unsuccessful species occurs at the species level where species themselves have the status of individuals. Because of structural and developmental constraints, the organisms, which are the individuals at a lower level in this particular hierarchy, are relatively limited in the extent to which the processes of adaptation can change them, and this limits the impact that such individual organism change can have on higher levels of the hierarchy. Whether a species is successful or not depends on species-level properties, such as population density and dispersal, rather than on the level of adaptations of individual organisms. Speciation is a species-level property. Adaptation is an organism-level property.

I do not know whether punctuated equilibrium theory will be absorbed into neo-Darwinism, or triumph and become the new and dominating theory of evolution. I am inclined towards the belief that neither will occur, and that a hierarchically expanded evolutionary theory will become more widely adopted in the future, with punctuationism and neo-Darwinism both being subsumed in a wider and more powerful theory. What I do know is that the puntuationists have been more successful than any previous theorists in bringing the idea of hierarchy to the centre of the theoretical stage. For this reason the last words of this section can be Gould's:

The issue is larger than the independence of macroevolution. It is not just macroevolution vs microevolution, but the question whether evolutionary theory itself must be reformulated as a hierarchical structure with several levels – of which macroevolution is but one – bound together by extensive feedback to be sure, but each with a legitimate independence. Genes, bodies, demes, species, clades are all legitimate individuals in some situations, and our linguistic habit of equating individuals with bodies is a convention only. Each kind of individual can be a unit of selection in its own right. Natural selection operating on bodies will not encompass all of evolution. Genes are the units of selection in the hypothesis of 'selfish DNA'; demes are units in Sewell Wright's shifting balance theory. Species represent one level among many; evolutionary theory needs this expansion.

THE CONCEPT OF ADAPTATION

Apples have a curiously ubiquitous place in our culture. From Adam and Eve to Newton via William Tell, with all that symbolism of apples being the fruit to give to the teacher when attempting to gain their favour, and how one a day keeps the doctor away. It must be some odd combination of appearance, availability and a texture which means they don't easily squash in school satchels, they can be cleanly cloven by an arrow, and they hurt when they fall on one. So it is that apples can be studied in a variety of ways – through their standing in folk-theories on health, honesty and inspiration, their place in the history of alcoholic beverages and their role in small economies. None of this, though, gets at what an evolutionary biologist would consider to be the essential appleness of apples. A physico-chemical analysis would not either – well, yes, chemically they are a touch different from other fruits, but not much. No, the appleness of apples lies in their unique structure: that shiny, waterproof skin, the undifferentiated pulp inside and at their centre those annoying pips and fibres.

This unique structure is related to what apples 'are for', that is, what the function of apples is: propagating apple trees. There is another way of putting this. Apples have a particular set of characteristics, a set of adaptations, whose goal is apple-tree propagation. It is this claim that establishes the differences between the biology of apples and all other ways of studying this fruit. Neither physics, nor chemistry, nor economics, nor mythology can explain apples. Only evolutionary biology, with its concept of adaptation, can do that.

This argument can be widened to the statement that the concept of adaptation at large is the central and essential claim for biology to be seen as an independent science. And it is very important to this book, because I will be arguing that knowledge as commonly conceived can only be understood in this same light. Essential to the understanding of knowledge is the general notion that it has evolved because it has a function, a goal or end-directedness. Towards the end of this book I will also argue that the processes by which it is gained in our heads, and the limitations on those processes, are the same as those that have given rise to the appleness of apples.

Adaptations are generally those features of plants or animals that astonish and delight us and lead us to exclaim at the intricate cleverness of nature. And both biologist and non-biologist alike share a particular way of thinking about adaptations. We look at the skin of an apple, the wing markings of a moth and the skeletal structure of a bird, and say that there is a reason for these attributes. The point, the goal, of the apple skin is to keep fluids from moving either in or out of the fruit. The goal of the markings on the wings of certain species of moth is to frighten away predators, which will be fooled into thinking that they are looking at the eye of a beast larger than they are. Birds have hollow, lightweight bones because such bones are lighter, and flight is made more easy. In each case we make the often unthinking assumption that these adaptations are 'for something', that they have goals, that they are end-directed. This habit of thought is so widespread as to make one think that we are all in some way predisposed to think that objects, at least living objects, in the world about us have purposes and goals. It certainly is an ancient way of thinking, because Aristotle formalized it in his theory of causes over 2,000 years ago.

Aristotle argued that any object in the world, animate or inanimate, has four types of causes and hence can be explained in four possible ways. These are the material cause (what a thing is made of), which corresponds roughly to the physico-chemical cause; the efficient cause (how something comes into being), in biological terms equivalent to development; the formal cause (the essence of a thing, what it is), which has no easy translation into modern scientific terms; and the final cause (what it exists for, its ends or goals), for which there was no scientific equivalent until the age of Darwinism. In natural objects, as opposed to those that are made by people, the latter three causes coincide, and so natural objects can be explained in terms of two over-arching causes, material and final. It is worth noting that Aristotle did not restrict the notion of final cause to living things. For him the entire universe is subject to a form-giving, finalistic force that drives all things towards a state of perfection. None the less, it is the living world that has been habitually seen as being in a state of perfected balance with the inanimate world about it.

Subsequently, the harmony and perfection of the living world were taken as evidence for the existence of a divine Being, of a supreme Creator who designed living things in accordance with the conditions of the world in which they exist. For more than 2,000 years the explanation of adaptations was ruled either by Aristotelian finalism or by notions of divine inspiration and creation – until the nineteenth century. What characterizes both is a 'forward-looking' principle, by which is meant that adaptations can be explained by some knowledge of the future which is either immanent in all things (Aristotelian finalism) or a part of God's wisdom. The form and coloration of stick insects were explained by a knowledge of the future of these insects, of the conditions of their future life when they were created, and hence a knowledge of the need for camouflaging features. The problem with both theological and Aristotelian explanations is that by placing the causes, that is the conditions of the creatures' life, behind the effects, which are the adaptations themselves, both these types of explanation are contra-causal. From Descartes and Bacon in the early seventeenth century onwards, such forms of explanation – sometimes grouped together and described as being finalistic or teleological (from the Greek *teleos*, meaning end) – have been rejected by scientists for just that reason. Science deals only with a posteriori causation where the cause precedes the effect; a priori causation, which is what teleological explanation is, is not acceptable.

One of the great achievements of both Lamarck and Darwin is that they provided scientifically acceptable a posteriori explanations of the causes of adaptations. Lamarck's explanation, as we know, was incorrect. Darwin's explanation, on the other hand, is considered in general terms to be correct. So let us consider how a finalistic interpretation compares with Darwin when dealing with the case of Little Red Riding Hood and the large, slashing teeth of her grandmother, who, of course, was actually a wolf. The reader may remember that, in response to Little Red Riding Hood's question as to why Grandma had such long teeth, the wolf replied, 'Because they are all the better to eat you with.' Now, there are two contrasting ways of understanding what the wolf meant by this answer. The finalistic interpretation is that the Creator, knowing that wolves will in the future be in a position to eat little girls, wisely equipped the beast with teeth admirably suited for slashing and cutting – or, in the spirit of Aristotelian finalism, the teeth of wolves are fashioned by the immanent knowledge of future encounters with little girls. In both cases, the explanation is one of a priori causation, the form of the teeth occurring prior to the appearance of little girls.

Darwin's explanation is quite different. Over aeons of their existence and evolution, wolves have been born with teeth of differing shapes and length. Little girls, and other similar prey, have been plentiful in wolf history. Those wolves with teeth most efficient in cutting and slashing survived more easily than, say, wolves with flat, blunt teeth. Tooth shape and length being heritable, the more successful wolves, which would include those with sharper and longer teeth, passed these traits on to their offspring. So, the wolf's teeth are admirably suited to the task of eating Little Red Riding Hood *now* because of the history of selection acting upon the variable and heritable trait of tooth form in the past. That means that the cause of the wolf's long teeth lies in the past history of variation and selection. This is a posteriori causation, and for that reason it is a scientifically acceptable explanation.

All adaptations are now explained in this general way, whether it be the exquisite structure of the human eye, the acceleration of a gazelle or the stealth of the tiger stalking the gazelle. Adaptations come in all forms – physiological, structural and behavioural. They are all the result of selection acting upon heritable variation, the most adaptive variants being those that either reduce the energy costs of the creature, or increase its energy intake, or in some other way enhance its likelihood of surviving and reproducing. Darwin's *Origin of Species* did away for all time with the problem of teleology.

There is one further aspect of every adaptation that must be understood. An adaptation is some form of organization of the phenotype relative to some feature of environmental order. Every adaptation has this dual characteristic of organismic organization and environmental order. It is precisely this relational quality of adaptations that gives them the appearance of being goal- or end-directed. So, for example, the stalking tiger possesses a set of behavioural adaptations that bear a relationship to a specific feature of the tiger's environment, such as the set of sensitive sense organs and evasive behaviours of gazelles – all, of course, in the context of the tiger's need for the energy resources that gazelles represent and the need of the gazelles themselves to survive and themselves reproduce. Since it seems to be psychologically so natural to think that the goal of the tiger's stalking behaviour is the capture of the gazelle, I don't see that there is too much harm done by framing the problem in this way. It is a convenient shorthand – provided that one is always ready and able to translate this easy way of putting things into the more stringent and exacting language both of the empirically accessible relationship between organismic organization and environmental order, on the one hand, and of the conceptually essential idea of any adaptation being the result of a long history of selection acting upon heritable variation, on the other. Incidentally, it is clear from my example that other organisms, indeed other members of the same species, can, and often do, represent the features of environmental order with which organismic organization is in adaptive relationship.

So all adaptations are the organismic end of such relationships, and none should be seen in isolation from the environmental factors that have provided the selection pressures for them. This relational quality of adaptations is the same as the relational quality of our knowledge, if the reader can recall the discussion of Chapter 1. In the case of knowledge as commonly understood, the relation is between a brain state and some feature of the world. In this restricted sense at least, and we shall steadily build on it, knowledge has the form of an adaptation.

Now, because it is the concept of adaptation that makes evolutionary biology different from all other sciences, and because the concept of adaptation placed the biological sciences on a more secure conceptual footing in the face of theological claims about the origins and causes of living forms, it has come under closer scrutiny than any other aspect of evolutionary theory. One result already mentioned is the dissociation that some evolutionists have attempted to establish between microevolution, which is the set of processes responsible for the formation of adaptations, and macroevolution. This is a very radical break with Darwinism, for that assumes a direct causal link between the adaptations of isolated breeding populations to local environmental conditions and the subsequent formation of separate species. More than that, it is reasonable to assert that for Darwin, and all modern Darwinists, adaptation to local conditions is the primary cause of the morphological and behavioural changes that might eventually lead to speciation. But this does not mean that adaptation is the only cause either of the changes in the phenotypes of populations over time or of speciation. Darwin himself was something of a pluralist in this regard, and so too are most neo-Darwinists.

Darwin recognized at least two other sources of change in phenotype, adaptive or otherwise. The one he referred to as 'correlations of growth', or what in modern parlance is known as allometric effects. Allometry is the study of correlations of growth and size. For instance, as animals get bigger, perhaps as an adaptation to predator pressure, all parts of their bodies increase in size, not just their bones and muscles. This sounds obvious, and it is, and so it is widely recognized as a cause of certain changes in body form or relation.

Brain size is an excellent example. Elephants have enormous brains in absolute terms, weighing in the region of 5,700 grams. That is more than four times the size of the human brain, which weighs in at an average of about 1,300 grams. Does this mean that elephant brains are big because the size is an adaptation to some aspect of elephant life, perhaps relating to their fabled powers of memory? Does it mean that elephants are giant walking memory banks? Well, no, they are not: we know of nothing that suggests that elephants are intellectually gifted in any way. However, we do know, of course, that elephants are big animals in every respect. An adult African elephant weighs about 6,700 kilograms, compared with the human adult average of around 65 kilograms.

Elephants have big brains because elephants have big everything. Their brain size is an allometric consequence of their body size. In fact, their brain size is very close to what one would expect it to be if one adjusts for body size. The human case is different. Our brain size is about seven times greater than would be expected from our body size. It is this difference, between what the brain size is in a species like *Homo sapiens* and what it would be expected to be on the basis of allometric effects, that needs to be considered in adaptational terms rather than the absolute brain size of large-brained beasts like the elephant.

Darwin was also aware of the possibility that a structure or behaviour might exist and serve a function now that is different from that for which it was originally selected. Having thus been recruited to some 'new' function, it may not constitute the perfect functional solution. A not dissimilar notion of 'exaptation' has recently been suggested, an exaptation being an adaptation that has recruited to functional requirement a phenotypic trait that either originated as a non-adaptive feature or first evolved for some other use. In this case too the adaptation might depart considerably from what an engineer would consider to be optimal design.

Neo-Darwinians recognize other reasons why adaptations need not be, indeed are most often likely not to be, perfect functional solutions to current demands. Adaptations take time to form and during that time the conditions of the world may change. Many adaptations may therefore lag behind the circumstances that the current holders of the adaptations are experiencing. (This is a crucial starting-point for the analysis of the evolution of learning and intelligence, which will be taken up in Chapter 5.) Also to be considered are the possible phenotypic effects of accumulation of neutral mutations and the consequences of evolutionary (molecular) driver mechanisms; the effects of structural and historical constraints; the limitations on 'perfection' imposed by restrictions on what genetical variants can be generated; and restrictions imposed by the actual building materials to hand. Indeed, if one takes all these factors into account, it is a remarkable testament to the powers of the microevolutionary processes that the adaptations seem so often to be impressive 'solutions' to the requirements of exploiting features of their environments – avoiding predation, consuming essentials such as water, and so on. The list of adaptive functions in any living creature is a long one.

Most adaptations, therefore, are a compromise between some perfect functional solution and what can, in fact, be achieved. This 'muddling through' view of adaptations is wonderfully well captured by the American H.A. Simon's neologism *satisficing*, which means the adoption not of optimal solutions but of practical, satisfactory ones. Very few biologists claim that every feature of an animal or plant is an adaptation; and few would claim that adaptations are optimal. Criticisms that have been levelled against evolutionary theory, often by evolutionists themselves, on these grounds are trivial, and often plain silly. So too is the criticism that evolutionary biologists are 'knee-jerk' adaptationists in that they look at an animal or plant and the first thing that they do is guess at what adaptive function some phenotypic structures might have; then when they look again, and find they were wrong, they guess again at some other adaptive function, and so on. Well, of course they do, but they do so in the awareness of factors such as allometric effects, multiple gene effects (known as pleiotropy) and the notion of constraint. The pejorative word 'guessing' should not blind one to what adaptationists are actually doing: it is called hypothesis testing, and it is normal scientific practice. However, my criticism of the critics does need to be balanced by the statement that *all* biologists, including the critics of knee-jerk adaptationism, accept the importance of adaptations. After all, if they did not, they would be denying the existence of evolutionary biology as a scientific discipline. It is worth repeating: adaptations, and the explanation of adaptations, are central to the biological enterprise, and that includes the social sciences. If we don't accept this, then biology becomes a reductionist exercise, and we will all be doing chemistry and physics.

There are, though, two criticisms levelled against adaptationist thinking that are correct and merit attention. The first is that, for reasons such as allometry and constraint just mentioned, it is not possible to take an animal, any animal, and to point with certainty to some of its features and declare them to be adaptations, and to point to others and say that those are not adaptations. Moreover, we don't know how to measure adaptations relative to different adaptations in the same animal or similar adaptations in other species. Now science is about measurement, and if we cannot measure the object of a science then is it really science? Well, yes it is, but it isn't yet a very advanced science.

These are serious deficiencies in evolutionary biology, and no amount of defensive huffing and puffing will gainsay these shortcomings. When we recognize an adaptation, it is through either a curious mix of intuition that this or that characteristic of an animal 'makes sense' relative to some feature of the environment, or a general sense that statistically unlikely complexity marks the presence of an adaptation. The streamlined body shape of most water-dwelling vertebrates, be they fish or mammals such as whales or dolphins, is easily connected in our minds to the requirements of body shape that are imposed by rapid movement through a relatively dense medium like water. Intuitions can, though, fail, as the best of adaptationists are quick to point out. A judgement on the grounds of complexity seems to be safer. The archetypal example here is the eye. It is inconceivable that an attribute as complex as the eye could be some accidental consequence of neutral mutations, complex gene actions, and the like. Richard Dawkins's recent book *The Blind Watchmaker* is a passionately argued and easily accessible account of how such complexity is explicable only in terms of the accumulated effects of microevolutionary processes. Despite such excellent accounts, it still is something of a scandal that we cannot measure adaptations when certain that that is what they are, and cannot always be certain in the case of less complex forms of organization that something is an adaptation at all. I am convinced that the answer lies in analysing the relational quality of adaptations rather than in seeing them in terms of their contribution to survival and reproduction, the latter being what biologists call 'fitness'. Many years ago the British biologist G. Sommerhoff provided the beginnings of just such an account, but it seems largely to have been ignored. If we resuscitated and looked again at his account of adaptive organization, we might be saved from the current weakness of adaptational theory.

The second criticism, which is owed largely to R.C. Lewontin, the American evolutionary biologist, is that we tend to take too static a view of the notion of adaptations; that we are too much inclined to see them as the solutions of passive creatures to the dominating forces of the environment. Yet it takes only a little thought to realize that living creatures change the world in which they live in a variety of ways. Trees cast shadows, dump leaf litter on the ground, suck water and nutrients out of the soil and exchange carbon dioxide for oxygen in the air. Trees don't just react to the world, they change it too, and so are in dynamic interaction with it. This is true of animals as well; the capacity of most animals for movement, if anything, increases the dynamics of the interaction. So adaptations are not static solutions to static problems. The dynamic nature of adaptations is pervasive, and is as characteristic of a particular set of adaptations called intelligence as it is of those concerned with, say, respiration. And again I believe that it is only by casting the analysis of adaptations into the context of their relational character, as Sommerhoff did, rather than by considering their contribution to fitness, that we can properly account for the dynamics of adaptations.

Fitness, as the American evolutionist G.C. Williams pointed out, is a statistical abstraction. Animals do or do not survive and reproduce, and each adaptation adds to mean phenotypic fitness. This is an elegantly economical view of adaptations, but it is not a dynamic one. By contrast, when an adaptation, some aspect of an animal's organization, works relative to some feature of its environment, it may have the effect of changing in some way that very feature, and that altered feature may in turn affect either that same adaptation or some other. The relational view of adaptations accommodates such dynamic interactions, whereas the fitness approach to adaptations does not.

Despite the legitimacy of such criticisms, I am convinced, as are most evolutionary biologists, that the concept of adaptation is at the heart of our science. As I will show in later chapters, it is also at the heart of a proper understanding of the nature of knowledge.

SUGGESTED READING

Dawkins, Richard (1986) *The Blind Watchmaker*, Harlow, Essex: Longman. A clear and simple exposition of neo-Darwinism, hostile to other theoretical approaches.

Eldredge, N. (1985) *Time Frames: The Rethinking of Darwinian Evolution and the Theory of Punctuated Equilibria*, New York: Simon & Schuster. One of the founders of punctuated equilibrium theory presents an alternative view to orthodoxy.

Ridley, Mark (1985) *The Problems of Evolution*, Oxford: Oxford University Press. Covers similar ground to Dawkins but is much shorter and more conventional in style.

3

Universal Darwinism

Speciation and adaptation are the two core phenomena that conventional evolutionary theory sets out to explain, though the relationship of these to each other varies depending upon which version of the theory one is following. Despite the title of Darwin's 1859 book, it is not uncommon for it to be said that Darwin actually seemed to be more concerned with the adaptive fit between the form and behaviour of living creatures and the conditions of the world in which they live, rather than with speciation as such. I don't think this is correct. Darwin saw an essential and ineradicable link between adaptation and speciation – what neo-Darwinists now refer to as the macroevolutionary consequences of microevolution. Lamarck, on the other hand, did not try to establish a link because he did not see it. His was literally a dual theory, concerned primarily with the explanation of adaptation through one set of ideas, and the transformation of species through another, unrelated set of ideas. There is a third route, one which contemporary punctuationists have gone down, which is to deny the necessity of the link between microevolution and macroevolution, but at the same time suggesting that the processes by which each occurs are identical.

This last position is a particularly interesting one. It is a view closely related to that of certain kinds of evolutionary epistemologists; and it is a view not a million miles away from that held by some very hard-nosed neo-Darwinians, though the latter argue fiercely for the retention of that essential link between adaptation and speciation. What unites these rather disparate theorists is adherence to the idea of universal Darwinism, the notion that the processes which drive the transformation of adaptations and/or species in time here on Earth have a rather wider existence than that depicted by conventional evolutionary theory. The English biologist Richard Dawkins, who was the first to use the phrase, meant by it that such processes are not bound to life on Earth but are the same wherever adaptation and/or speciation are occurring or have occurred anywhere in the universe. This is a very bold and interesting claim but, alas, not one which for the foreseeable future can be put to the test. It is going to be many years before extraterrestrial life forms can be examined for the presence of Darwinian processes.

There is, though, another meaning of the phrase, namely that it isn't only adaptation and speciation as conventionally understood that are driven by these processes, but that here on Earth certain other forms of transformation of living systems are also caused by these same processes. This is a more immediately testable extension of evolutionary theory into universal Darwinism, because we can look inside our own bodies and those of other creatures now to ascertain whether there are indeed processes embodied in 'Darwin machines' that conform to Darwinism. So, while duly acknowledging Dawkins's precedence in using the phrase, I will confine it here to the assertion that there exist processes *inside* of organisms that conform to Darwinian evolution as well as the more conventionally understood processes that operate *between* organisms.

This distinction between evolution *within* as opposed to evolution *between* is so easy to confuse and misunderstand that it is worth dwelling on the difference. As conventionally understood, evolution describes the processes by which certain forces act to differentiate between, and then differentially to propagate, whole organisms. This is evolution *between* organisms. Evolution *within* concerns the processes by which certain forces act to differentiate between, and then to differentially propagate, components or states or parts of organisms, but only during the lifetime of those individual organisms. Because of this difference, evolution *between* operates on time scales well in excess of that of the lifetime of single organisms, whereas evolution *within* involves time scales less than the lifetime of the single organisms within which such processes occur. What is meant by universal Darwinism here is that both forms of evolution, between and within organisms, conform to identical processes.

If this form of universal Darwinism is correct, then it might have very important implications for phenomena that obviously go far beyond the confines of conventional evolutionary theory, and the latter might then be reduced to just one aspect of a much wider theory. This chapter will consider the possible forms that universal Darwinism might take, and explores what the concept really means.

A VERY BRIEF EXCURSION INTO HISTORY

The beginnings of universal Darwinism go back almost to the beginnings of Darwinism itself. In a book review published in 1869, Darwin's friend and defender T.H. Huxley suggested that it might be useful to extend the ideas of a Darwinian struggle for existence between organisms to the struggle for existence within an organism where the struggle might be thought of as occurring between its developing parts and constituents. Some of the parallels that have been drawn between Darwinian natural selection theory and how other biological systems and phenomena might work are so strange and unexpected to conventional evolutionists, but also so vivid, that over the next few pages I will quote directly from their writings, including some of the most famous evolutionists of all – they make the point much better than I could. For example, Huxley wrote:

It is a probable hypothesis that what the world is to organisms in general, each organism is to the molecules of which it is composed. Multitudes of these, having diverse tendencies, are competing with one another for opportunity to exist and multiply; and the organism as a whole is as much the product of the molecules which are victorious as the Fauna and Flora of a country are the product of the glorious organic beings in it.

Darwin himself was ambivalent in his attitude to this early attempt to expand Darwinism in this way. In that same year, 1869, he wrote a letter to Huxley in which he said: 'I am very glad that you have been bold enough to give your idea about Natural Selection amongst the molecules, though I cannot quite follow you.' However, in *The Descent of Man*, published just two years after he had written that letter to Huxley, and hence the writing of which must have overlapped with the correspondence quoted above, Darwin wrote of language development in terms couched so strongly within an extension of the theory of natural selection that it is worth quoting at some length:

The formation of different languages and of distinct species, and the proofs that both have been developed through a gradual process, are curiously parallel. But we can trace the formation of many words further back than that of species, for we can perceive how they actually arose from the imitation of various sounds. We find in distinct languages striking homologies due to community of descent, and analogies due to a similar process of formation. The manner in which certain letters or sounds change when others change is very like correlated growth. We have in both cases the reduplication of parts, the effects of long-continued use, and so forth. The frequent presence of rudiments, both in languages and in species, is still more remarkable. The letter *m* in the word *am*, means I; so that in the expression *I am*, a superfluous and useless rudiment has been retained. In the spelling also of words, letters often remain as the rudiments of ancient forms of pronunciation. Languages, like organic beings, can be classed in groups under groups; and they can be classed either naturally according to descent, or artificially by other characters. Dominant languages and dialects spread widely, and lead to the gradual extinction of other tongues. A language, like a species, when once extinct, never, as Sir C. Lyell remarks, reappears. The same language never has two birth-places. Distinct languages may be crossed or blended together. We see variability in every tongue, and new words are continually cropping up; but as there is a limit to the powers of the memory, single words, like whole languages, gradually become extinct. As Max Muller has well remarked: 'A struggle for life is constantly going on amongst the words and grammatical forms in each language. The better, the shorter, the easier forms are constantly gaining the upper hand, and they owe their success to their own inherent virtue.' To these more important causes of the survival of certain words, mere novelty and fashion may be added; for there is in the mind of man a strong love for slight changes in all things. *The survival or preservation of certain favoured words in the struggle for existence is natural selection* [emphasis added].

Darwin may not have felt able to follow Huxley in an extension of the theory of natural selection to molecules, but he clearly had no inhibitions about a quite detailed speculative foray into the role of those same evolutionary processes that lead to speciation in the development of language, both within individuals and in communities of language users. In the 1880s there were others who followed Huxley's lead when writing about development, and Huxley himself in 1881 extended to thought the idea that evolutionary processes might be operating within organisms: 'The struggle for existence holds as much in the intellectual as in the physical world. A theory is a species of thinking, and its right to exist is coextensive with its power of resisting extinction by its rivals.' Clearly Huxley envisaged very wide applicability for the principles of competition and selection. But it was the American philosopher and psychologist William James who, a year earlier in an essay entitled 'Great men, great thoughts, and the environment', first extended this idea to the discussion of learning, thinking and culture. James began the piece by noting that 'A remarkable parallel, which to my knowledge has never been noticed, obtains between the facts of social evolution and the mental growth of the race, on the one hand, and of zoological evolution, as expounded by Mr Darwin, on the other.' James was scathingly critical of the English philosopher Herbert Spencer's psychology, which denied 'the vital importance of individual initiative' in the production of thoughts and ideas. He provided an extended argument against Spencer's Lamarckianism and in favour of the Darwinian, and hence undirected, nature of evolution, undirected because the variant forms are unrelated to selection pressures. James concluded:

I can easily show that throughout the whole extent of these mental departments which are highest, which are most characteristically human, Spencer's law is violated at every step; and that, as a matter of fact, the new conceptions, emotions, and active tendencies which evolve are originally *produced* in the shape of random images, fancies, accidental outbursts of spontaneous variation in the functional activity of the excessively unstable human brain, which the outer environment simply confirms or refutes, adopts or rejects, preserves or destroys – *selects*, in short, just as it selects morphological and social variations due to molecular accidents of an analogous sort [emphasis in the original].

One must be clear about what James is saying here. His main point is that original ideas are not impressed upon the originator by forces in the external world. Rather, they begin as random variations that have their source somewhere in the human thought mechanism. When *first* produced, these random products are independent of all external events. Once produced, they may be selected, and that process of selection certainly does involve events external to the devices that originally produced the variation. That is why James was so scathing about Spencer's Lamarckianism, which he judged essentially deprived people of the capacity for original thought and reduced them to machines upon which experiences and understanding are forced by the outside world. But for an idea to be original it has somehow to come from inside the originator rather than merely be imprinted by external circumstances. Originality and creativity are, after all, defined by the independence of their products from direct experience. Hence James's stress on 'individual initiative'.

James explicitly denied that such internalized evolutionary processes could be applied to the 'lower strata' of the mind, including the 'entire field of habit and association by contiguity', which he held to be 'passively plastic', and hence slavish and unoriginal in its products. In the terms of selection and instruction described in the previous chapter, James considered that creative thought occurs by a process of Darwinian selection, whereas associative learning is the result of Lamarckian instruction. It was the American psychologist James Mark Baldwin, writing around the turn of the century, who specifically extended the notion of Darwinian selection theory to associative learning. 'The individual's learning processes are by a method of "trial and error" which illustrates "natural" in the form of "functional selection",' wrote Baldwin, whose interests, however, extended far beyond simple learning. He was particularly interested in behavioural development, and in the way in which the development of the individual relates to evolution. To this end he formulated the concept of organic selection. The following quotation is from a 1909 publication:

If, that is, a selection of processes and habits goes on within the organism – a functional selection resulting in a real molding of the individual – there would be at every stage of growth *a combination of congenital characters with acquired modifications*: natural selection would fall in each case upon *this joint or correlated result*; and the organisms showing the most effective combinations would survive. *Variation plus modification*, the joint product actually present at the time the struggle comes on, *this is what selection proceeds upon*, and not, as strict neo-Darwinism or Weismannism supposes, upon the congenital variations taken alone. The result is that variation would tell *most when in the direction in which the accommodations were being made and found useful*; and on the other hand, accommodations would be made *where variations best permitted*. There would then be an accumulation of variations, 'coincident' in direction with the acquired modifications, the function becoming more and more congenital from generation to generation. The accommodations and modifications of the individual serve as a supplement or screen to his endowment; and in course of time the endowment factor, by variation simply, with no resort to the actual inheritance of acquired characters, comes to its perfection. This result of the 'coincidence' of modification and variation in guiding the course of evolution has been called 'organic selection' [emphasis in the original].

Organic selection is not an easy idea to understand, but it is worth dwelling on as it concerns the relationship between evolution and individual development, which is an old and contentious issue in biology. What Baldwin was saying is that selection does not operate upon genetical variation alone but upon phenotypic variation which is the outcome of both genetically and developmentally generated variation. His point was that variation is importantly determined by within-organism events. Moreover, he envisaged the actions of genetical and developmental sources of variation as acting in concert to result, over numbers of generations, in the selection of developmental pathways that, though initially largely (never only) environmentally determined, come increasingly to be caused solely by genetical means. This claim makes organic selection, despite the protestations of Baldwin's followers to the contrary, a notion lying on the borderline of Lamarckianism (in the sense of the inheritance of acquired characters).

However, whatever the worth of the idea of organic selection as a form of internal evolutionary process, empirically there *is* a well-known phenomenon called genetic assimilation, first demonstrated by an English biologist, C.H. Waddington, long after Baldwin was dead, which does seem to support at least part of Baldwin's formulation. 'Canalization' is the word that Waddington gave to a particular manifestation of genetic assimilation and refers to the way certain developmental pathways become buttressed by the accumulation of appropriate genetic instructions. Canalization, which Waddington argued is the result of an emphatically Darwinian process, is an idea which is seemingly, if not actually, in direct line of conceptual descent from Baldwin's organic selection.

Caught up in a considerable personal scandal, Baldwin had to leave his job and his country; after 1909 he lived in France and for that reason he subsequently had a greater effect upon European scientists than he did upon his North American colleagues. Specifically, his writings influenced Jean Piaget, one of the fathers of European structuralism, who for many decades argued for a close interrelationship between evolution and cognitive development. This radical brand of evolutionary epistemology was labelled genetic epistemology by Piaget. I dismissed structuralism as a form of evolutionary theory in the last chapter as mere hand-waving. Worse, Piaget's genetic epistemology came eventually to advocate Lamarckianism in biological evolution. I therefore don't propose to spend much time on an incorrect theoretical position. But Piaget has had a great impact on twentieth-century psychology, and for that reason it is worth considering the reason for his radicalism and his error.

Piaget was an evolutionary epistemologist in that he believed that cognition is made up of brain processes that are 'an extension' of fundamental evolutionary processes. In other words, he subscribed to the form of universal Darwinism advocated here, which is that there are within-organism evolutionary processes that are of the same nature as the between-organism evolutionary processes. However, it was with the conventional formulation of such between-organism processes that Piaget was at odds. He viewed with great scepticism the notion that chance events lie at the centre of the evolutionary process. Specifically, he rejected the conception that chance variations (mutations) are the ultimate source of new genetic material, that changes in genes are unrelated to the selection pressures acting on the organism at the time that the new genetic material is formed, and that chance segregation and cross-over effects in the formation of sex cells are important in disseminating genetic changes through breeding populations. He also envisaged a crucial role for behaviour in evolution:

either chance and mutation can explain everything or else behaviour is the motor of evolution. The choice is between an alarming waste in the shape of multitudinous and fruitless trials preceding any success no matter how modest, and a dynamics [sic] with an internal logic deriving from the general characteristics of organization and self-regulation peculiar to all living things.

The Piagetian conception of behaviour as an evolutionary 'motor' will be discussed in Chapter 6. The important point here is that for Piaget the identity between cognitive processes and all other processes that establish and maintain the organization of living things lies not in the chance-variation-and-selective-retention model, but in their all being adaptations (adaptations, including adaptive cognitive structures, being characterized by their organiza-tion, according to Piaget). He noted that 'the constant functional conditions of the process [of adaptation] are two in number – assimilation and accommodation'. Assimilation for Piaget meant the incorporation of the external world into cognitive structures, but the requirements for the incorporation of such representations into already existing cognitive structures impose necessary changes upon the way the world is represented. Accommodation is the ensuing change in cognitive structure that occurs as it adjusts in order to fit the representation of the conditions of that external world. All adaptations are the outcome of a dynamic, a dialectic, and endless interaction between assimilation and accommodation, and both are present at all levels of adaptive organization –

genetic, epigenetic (developmental), individual learning and logico-mathematical thought. So it is that Piaget's genetic epistemology centres around the structuralist ideas of organization, self-regulation, co-ordination and construction rather than the conventional evolutionary concepts of variation and selection.

Piaget's fame rests on his developmental psychology, and this raises the issue of the current status of the idea that development of any kind, not just psychological or behavioural, might be at least in part understood in terms of within-organism evolutionary processes. This takes us back to where it all started, with T.H. Huxley's suggestion of the extension of the principle of selection to individual development. Just recently there have been renewed attempts to look again at the extraordinary transformation from a single fertilized egg to a functioning multicellular organism and see whether internalized evolutionary processes cannot be discovered. An over-proliferation of cells and cell lineages early in development does seem to be followed by a reduction, perhaps by selectional processes, to a restricted number of cell lines and forms. Whether real selectional mechanisms will be discovered in the development of all organ systems we will have to wait and see, but descriptively there are many known instances of development that do not appear to conform to some simple unfolding of genetical instructions. Most development occurs through the action of highly complex, active and interactive processes. It is likely that internalized evolutionary processes are among them.

There is one organ system, the brain, the development of which, as well as its post-developmental functioning, can certainly be viewed as conforming to an evolutionary process. This returns the historical story to 'mainstream' evolutionary epistemology of the William James and James Mark Baldwin variety, the most influential recent exponent of which is Donald Campbell, the American psychologist already referred to in Chapter 1. Since 1974, when Campbell first used the phrase 'evolutionary epistemology', there has been a marked increase in work in this area, including some detailed theorizing about the neurophysiological processes underlying perception, learning and memory. I shall have more to say in a later section of this chapter about Campbell's contribution to universal Darwinism; and part of Chapters 5 and 6 will be a general exposition of this modern line of evolutionary epistemological thought. But the historical story is complicated, and for the moment I want to move away from brain function and look at two other strands in the development of these ideas.

Related to Campbellian evolutionary epistemology is its extension to the way science itself should be seen as a set of Darwinian evolutionary processes. For many years Sir Karl Popper, a philosopher of science, has advocated a particular view of how science should be understood as working, and on the basis of which scientists themselves should consciously establish their own activity. Popper argues that all of science is a process of conjecture and refutation, and that this indeed is what demarcates science from other ways of trying to understand the world. Scientists, he says, have ideas or, to be a bit more formal, hypotheses about some aspect of the world – say, the molecular configuration of a protein or the effects of spin of a subatomic particle – which then explain certain things about the world. These conjectures or hypotheses are then tested against the real world by observation and experimentation. Those conjectures that are found on testing to be wrong will then be discarded. The remaining conjectures will be refined and multiplied as new conjectures, and these in turn will be subject to tests, which will form the basis of further rejection and refinement. In this way science itself can be seen to be an evolutionary process, the product of a 'Darwin machine'.

As Popper himself stresses, his view is not meant merely to be understood metaphorically. He means it literally:

All this may be expressed by saying that the growth of our knowledge is the result of a process closely resembling what Darwin called 'natural selection'; that is, *the natural selection of hypotheses*: our knowledge consists, at every moment, of those hypotheses which have shown their (comparative) fitness by surviving so far in their struggle for existence; a competitive struggle which eliminates those hypotheses which are unfit . . . This statement of the situation is meant to describe how knowledge really grows. It is not meant metaphorically, though of course it makes use of metaphors. The theory of knowledge which I propose is a largely Darwinian theory of the growth of knowledge. From the amoeba to Einstein, the growth of knowledge is always the same: we try to solve our problems, and to obtain, by a process of elimination, something approaching adequacy in our tentative solutions [emphasis in the original].

Science itself is a particular form of culture; and again, the idea that culture itself changes by evolutionary processes is an issue to which we will return later in the book. But I hope that by now the reader will have a sense of the extent to which the notion that evolution is a process that can explain more than speciation and adaptation as traditionally understood has been used over the last 130 years and more. In addition to the names of those already mentioned in this section can be added Georg Simmel (the sociologist), Ernst Mach (the physicist), Konrad Lorenz (founding father of ethology and Nobel prize-winner), B.F. Skinner (the behaviourist psychologist) and H.A. Simon (social scientist, polymath and another Nobel laureate, this time in economics); there have been others too. Obviously it would be an error to consider the notion that evolution is something that occurs within as well as between organisms to be a recondite intellectual thread running in a barely detectable way through the writings of obscure scholars. But there is one case in the history of the line of thought that I am calling universal Darwinism that is so important to the idea that evolution may occur within as well as between organisms that it deserves rather closer attention than any other topic in this section. It involves an organ system, the vertebrate immune system, which is now known with certainty to operate through Darwinian processes.

Almost all living creatures have ways of dealing with substances, foreign bodies, that may physically enter them, as occurs most typically in the case of infectious diseases caused by such invading micro-organisms as bacteria and viruses. In animals these include actual physical capture and destruction of the invaders as well as more subtle chemical and detoxification devices. In vertebrates, including mammals like ourselves, this includes what is known as the immune system, which is closely involved with the blood system.

Blood cells and immune system cells are manufactured in the same parts of the body; they both circulate freely through tissues and the blood vessels; and the immune system has its own complex vasculature, known as the lymphatic system, through which circulate the cells and fluids of that system. There are several different kinds of cell in the immune system that protect the body against foreign substances and organisms. There are those that are non-specific in action, some of which are phagocytic, literally engulfing and destroying bacteria or the remnants of damaged cells; some produce enzymes that detoxify foreign substances, especially proteins; others exude chemicals, such as histamine, that aid in the general combating of foreign substances. Then there are very important kinds of cells, known as lymphocytes, whose role is to manufacture the antibodies that inactivate and destroy specific invading micro-organisms and their products, which are known as antigens.

The immunological response of antibody production which confers immunity to specific diseases has been known since Edward Jenner, an English physician, first described the immune response to smallpox in the 1790s. Louis Pasteur in France made further vital discoveries in the nineteenth century about the immune response to diseases such as anthrax. The immune system is absolutely vital to our well-being and survival. We know, from various forms of disease that damage or destroy it, that without a properly functioning immune system we are rapidly overwhelmed by all manner of illnesses that present no serious difficulty to people with normal immune system function.

How does the antigen-specific immune system work? The understanding of the immune system is one of the triumphs of modern molecular biology, and the answer can only be given here in most general terms. This will do some violence to a depiction of the functioning of an organ system that is at least as complex as the nervous system. Indeed, there are about ten times more lymphocytes in our bodies than there are nerve cells, and these lymphocytes are manufactured in or circulate through bone marrow, spleen, lymph nodes, thymus and appendix, as well as general blood and lymph vessels. So, like the nervous system, the immune system is made up of very large numbers of cells, which demonstrate great diversity of structure, and which have a widely dispersed distribution throughout the body. Now, until the 1950s, the immune system was thought to acquire immunity to disease through a Lamarckian instructional mechanism. That is, it was thought that when antigen enters an animal the antigen serves as a template to direct the synthesis of antibodies. In other words, it was thought that antibodies are produced as a directed adaptive response to the needs of the animal, with lymphocytes having their highly malleable structure moulded into specific antibody form by the structure of the antigen.

Modern immunology dates from the work of N.K. Jerne in Europe and F.M. Burnet in Australia, who advocated a quite different theory of immune system function known as clonal selection theory. It states that the immune system operates by the production of an enormous diversity of lymphocytes, some restricted sample of which is *selected* by fit to an antigen; the selected lymphocytes are then clonally reproduced and propagated into the future as antibodies. It is now known with certainty that clonal selection theory is the correct explanation of how immunity to disease is acquired. As a process of transformation in time, the population of lymphocytes being the entity that is changing, it conforms to a selectional evolutionary process in that it involves successive stages of over-proliferation of diverse forms (the highly variable lymphocytes), a testing or selection phase during which antigen is matched to specific lymphocytes, and then the proliferation and propagation of lymphocytes as antibodies.

There are no ifs and buts about how, in general terms, the immune system works. It is a 'Darwin machine' – an organ system whose transformation through successive adaptational states in time is explained by a Darwinian evolutionary process. The immune system does not work *like* an evolutionary process. Immune system function *is* an evolutionary process.

A FANTASTIC TALE

By now the reader should be starting to get a feel for universal Darwinism. I want to consolidate by indulging in a little excursion into fantasy. This will be a game of 'what might have happened had Darwin worked in other areas of science'. Later I will convert the fantasy into a consideration of the processes that are general to those areas of science in which our fictional Darwin worked.

As everyone knows, Darwin was a naturalist. For all of the reasons outlined in the previous chapter – including the increasing understanding of the prodigality of nature with regard to species diversity, the growing awareness of the great age of the Earth and the gains in evidence on the distribution of species in time (as evidenced by the fossil record) and space – the Darwin–Wallace discovery of the principle of natural selection was bound to have been directed towards the origin of species. None the less a whimsical flight of fancy and a little time-travelling allow us to reflect on what might have happened had Darwin been a psychologist, an immunologist or a scientist of science itself.

We can begin the fantasy on the firm ground of some of the very earliest laboratory experiments on learning, carried out in the United States at the turn of the century by one Edward L. Thorndike. He placed a cat in a box with a string hanging from the ceiling. The box was so constructed that a tug on the string would release the latch securing the door. And outside of the box, in full view and sniffing range of the cat, was a highly desirable and odoriferous piece of fish. Thorndike observed when he first placed the cat in the box that it displayed quite high levels of seemingly random behaviour as it tried to find a way of escaping from the box and reaching the fish. The cat tried all sorts of things like pawing at the walls, pushing into corners or rearing upwards. Occasionally, and quite by chance, the animal touched the string. Eventually it grabbed the string in both paws and pulled, not because it knew that this would release the latch, but because it is just another cat-like bit of behaviour. Of course the door did indeed open, allowing the cat to leave the box and consume the fish. Next day Thorndike put the cat back in the box with another piece of fish on the outside. Again the cat displayed a range of different behaviours until again, perhaps sooner than on the first day, it contacted the string, pulled on it and reached the fish. Thorndike repeated the procedure with a number of cats over many days and found a similar pattern in all of them. With repeated confinement in the box the cats came to pull the string sooner and sooner; eventually each cat would pull the string and get to the fish as soon as it was placed in the box.

Anyone familiar with cats will, of course, find the story wholly unremarkable, but as will be pointed out in the next chapter, this was, amazingly, one of the first laboratory demonstrations of learning in an animal other than a human. What is important here is Thorndike's interpretation of the pattern of his findings. Obviously the cats began the experiment not knowing how to release the door latch. They began their attempt to escape by doing a variety of things, which included tugging at the string. How soon they would first pull on the string was determined by some mixture of chance, the prominence of the string and the tendency of cats to play with objects like string. Gradually the random behaviour gave way to more directed string-pulling as the cats temporarily eliminated from their immediate behavioural repertoire those responses that did not lead to the door opening. This trial-and-error form of learning, also known as instrumental or operant learning, is widespread among vertebrate animals, including ourselves.

Think of how you yourself would go about opening an unfamiliar door. Of course, you would bring to the exercise a past history of door opening and so would know much more than cats to begin with. None the less, though directed at the door, your own behaviour would be a matter of random tugging, pushing and fiddling with bolts and catches until you hit upon the right sequence of behaviour. You would then, when faced with that same door in the future, eliminate the responses that had not worked before and concentrate on what did. So learning of this kind is characterized by a reduction in initially random behaviour that leaves in place, and possibly strengthens, behaviour that leads to rewards or avoids punishment – or what psychologists call reinforcements. A reinforcement (or a reinforcer), in general terms, is almost any event of consequence for an animal, one that the animal does or does not like and towards which it is not neutral in its response. Food, cold, a mate – rewards and punishments, in short – are all examples of reinforcers.

The trial-and-error learning experiment just described concerned changes in a relatively restricted set of behaviours relating to two features of the cat's world, the string and the fish. However, it is possible to expand the scenario to take in a much wider range of behaviours and many aspects of the animal's world, which is something that Darwin as a psychologist with a particular interest in animal learning might have done. In his learning laboratory he placed his subjects, small furry creatures, in a large open field so structured that certain behaviours have certain consequences, meaning that they may lead to reinforcement. Some of these are pleasant and sought after by the animals. Other consequences are unpleasant and avoided. Over time the features of the environment in the open field are changed, and so too are the consequences of some of the behaviours of the animals. Darwin, of course, was an excellent observer. He carefully noted down everything that the animals did, and he realized that he could quantify the frequencies of different behaviours and calculate their changing probabilities of occurrence. Most importantly, he also noted that, during periods characterized by nothing of consequence occurring to the animals, they displayed a great deal of diverse, undirected behaviours, including novel ones not previously observed.

When events of consequence, that is, things that the animals do or do not like and towards which they are not neutral in their response, do occur, some of these behaviours are associated with these consequences in time, and certain features of the environment that afford these behaviours are associated in space with these consequences. Darwin noted that following these events there is a significant change in the probabilities of the behaviours making up the animals' behavioural repertoire. Some increase in frequency, while others decrease in frequency or disappear; those whose probabilities of occurrence change most are those most closely associated with these events of consequence.

Darwin concluded that his animals somehow have the capacity for generating diverse behaviours, though he has no idea as to what the mechanisms are that underlie this capacity. He also concluded that the animals have the ability to narrow down this diversity by some process of internal selection by which behaviours that 'resulted' in good consequences increase in probability and those that led to bad consequences decrease in frequency. This reduction in diversity means that only certain behaviours are carried forward in time, and that they then come to be mingled into a behavioural repertoire of increasing diversity again until another phase of diversity reduction occurs. Of course, he didn't understand at all why some events are 'good' or perceived as 'good' by his animals, while other events are 'bad' or perceived as being 'bad'. Future investigators would have to sort out that puzzle. But he believed that he could formulate a theory of behavioural transformation in time, the changes resulting in behaviours that fit better with the current environment than other behaviours.

And so Darwin wrote a book entitled *The Origins of Behaviours by Means of Internal Selection or the Preservation of Favoured Acts in the Struggle for Being Remembered*. A theory of evolution is born which happens to centre on behaviour and behavioural change. Neuroscientists are subsequently able to explain the neural mechanisms that generate the diverse behaviours and why it is that certain events, reinforcers in the jargon of learning theory, change the probabilities of associated behaviours. Later generations of biologists with a special interest in the 'problem of species' realize that Darwin's theory of behavioural evolution may have something to tell them about the origin of species. One of them even writes a fantastic tale in which Darwin, had he been a naturalist, might have entitled his book *The Origin of Species by Means of Natural Selection or the Preservation of Favoured Races in the Struggle for Life*.

What of Darwin the immunologist? He knew that the immune system produces defender cells with membrane characteristics of quite staggering diversity. He had, at one stage, thought that there was much less diversity and that the defenders were almost infinitely malleable and able to change their structure as a function of the characteristics of foreign invaders, the change in structure then enabling the defenders to disarm and destroy the intruders. His mentor, a fellow immunologist by the name of Lamarck, was the originator of this idea. It was a notion of guided or instructed change, where the environment directs the changes in the defending cells. But Darwin later realized that the evidence for undirected diversity prior to the appearance of intruders was overwhelmingly strong, and that there was no empirical support for directed change as the basis for the way the immune system works. Again, he did not understand the source of the diversity, but no matter. What he did know is that following invasion by a foreign substance or organism, the latter is 'engaged' by a defender with the appropriate cellular structure, and that defenders that are thus able to 'fit' their own structure to that of the invader increase in frequency. That is, the fit between defender and invader initiates a selectional process whereby clones of the defender are produced in large numbers in order to destroy other invaders of the same or similar kind. Darwin calculated the probabilities, and his sums show that the probability of defenders of certain characteristics increases as a function of the excellence of their fit to the invaders. In this case his book is entitled *A Selection Theory of the Immune System* (with apologies to Burnet, who published a book in 1959 entitled *The Clonal Selection Theory of Acquired Immunity*).

Wearing his hat as a student of how science works, Darwin noted that every science goes through periods of being in the doldrums, when people experiment and observe widely and talk a great deal, producing a plethora of ideas. He observed the effects of the discovery of novel phenomena, the development of new methods and the not inconsiderable part played by the social structure of science in terms of who knows and likes and is indebted to whom and hence who cites whom. He charted the way in which diverse forms of empiricism and theory may become narrowed by the views of individual scientists arguing that certain lines of research are better formulations and explanations of particular phenomena than others. Darwin realized that science is a process of transformation and succession whereby periods of great diversity of work are interspersed with times of work of much narrower focus due to the judgements that certain ideas or methods or research programmes produce data and theory that fit better with the real nature of the world. Here too he realized that much work is needed before the precise sources of diversity will be understood, and that the kinds of judgements and social interactions determining the selection of some ideas and methods over others have yet to be determined. But he did recognize that, even in the absence of these details of mechanism, he had discovered a set of processes that together explain the changes in scientific activity and thought.

Yes, it is a fantastic tale, but the point of such fantasies must now be clear. It is that evolution as we usually understand it need not be confined to speciation–adaptation and changes in gene-frequency. For all the obvious historical reasons that our understanding of evolution should be concentrated around these issues, there is a real sense in which evolution as a process might have been discovered and formulated had Darwin been any of the kinds of scientist portrayed above. I want now to characterize more explicitly that general, mechanism-independent process which is at the heart of every 'Darwin machine', driving evolution wherever it occurs, and then to consider what mechanisms, what entities, might embody such a process. As a way of introducing these abstract ideas we can begin with the rather general consideration of how all of science works.

GENERAL PRINCIPLES, LAWS AND MECHANISMS IN SCIENCE

One of the objectives of science is to explain more and more with less and less. By this I mean that scientists hope that ever-greater numbers of phenomena in the world can be understood by the smallest possible number of scientific laws or principles. The discovery and articulation of these laws are, therefore, one of the primary responsibilities of science. What these laws or principles do is describe the deep regularities of the world. But these regularities are not 'just there'. They are themselves caused; they are the result of the characteristics of particular 'bits' of the world. So what could be considered as another aim of science, but is really an accompaniment to, and consequence of, the discovery of the laws of nature, is the identifying of those 'bits' of the world that embody and cause the regularities.

The simplest way to say all this is that the laws or principles of nature are an abstract description of these regularities, whereas their embodiment describes those bits of the world that provide the actual causal mechanism of these regularities. Now, these mechanisms are 'things'. We would be able, if our fingers were the right size, to touch and manipulate them. This 'thinginess' of mechanism means, in effect, that science aims to know *what* it must touch in order to identify the causal mechanisms of nature's laws. And this involves the drawing of conceptual divisions of the world so that the component bits that are significant can be distinguished from those that are trivial, because given the right kinds of fingers we could touch anything. The trick is to know what to touch. This is what Plato meant by his aphorism that science is the business of carving nature at its joints.

This issue of distinguishing the significant from the insignificant bits is so important to evolutionary theory that it is worth lingering over in rather general terms. Consider what I see if I look out of my study window. There are houses, gates, roads, cars, trees, shrubs, clouds and innumerable other objects. These are all of such a scale and familiarity that I notice them at once. If I used binoculars I would be able to add greatly to the list. Now, how, of the very many possible ways, could I best go about partitioning, classifying and measuring the world that I can see from my window in a way that is significant for our understanding of it? Little thought is needed to realize that there is, at present, no *one* way of doing this that is better than any other. It depends on what it is that I want to understand. For example, one form of understanding is through the science of cognitive psychology, and a psychologist might proceed to partition the world in terms of scale or familiarity: the significant objects would be things like trees and cars and houses. By contrast, a physicist, though living in houses and driving about in cars and subject to the same psychological forces that make scale and familiarity as significant to him or her as they are to the rest of us, none the less would not consider either of these of much interest to his or her physics. Physicists are much more interested in very small objects, so small that we need extremely large magnification to detect them at all. Other kinds of scientists might want to lump certain objects together into complexes of significant things, as for instance an ecologist might do with trees and clouds, because of the connections that ecologists make between these kinds of objects.

So what constitutes the 'best' way of dividing up the world, what the significant things are that one would want to touch in order to identify the causes of the deep regularities of the world, will vary depending upon the kind of scientist one is. Eventually, when we understand everything about the world, we will have a grand unified theory of everything and will want to divide the world up in only one way. But that is some time off in the future. For the moment most sciences seek to establish regularities relative to their own special view of the world and to divide the world into important and unimportant parts in their own particular way.

The search for the general laws or principles of nature and their embodiment in significant bits of the world or universe is exemplified by the history of physics in this century, which is the history of attempts to understand the relationships between the fundamental forces of nature like gravity, the binding forces of the atom and the electrochemical force of electrical and chemical reactions. The expression of these relationships is the laws and principles of physics, such as the inverse square law that describes the gravitational force acting between two objects. The actual 'thinginess' of nature that causes these forces – that is, what the most basic and irreducible bits of matter are that underlie these forces – is the quarks, strings or superstrings (or some as-yet-to-be-conjectured entity) that embody these forces in their particular characteristics. So the quarks or strings are the important bits of the world that the physicist wants to look at, put his or her finger on, so to speak, and measure. The overall conceptual result is that, now armed with this minimum number of forces and entities, physicists are beginning to understand very grand events like the origins of the universe, as well as being able to explain more mundane matters like the conversion of chemical to kinetic energy (which tells us how a stationary car with a chemical fuel in its tank can, via the controlled mechanism of the internal combustion engine, become a moving car with less chemical fuel in its tank). The range of explanation using so few concepts is extraordinary, and this is what makes physics so elegant and powerful as science.

There is another general point to be made through the example of physics that is important enough to bear repeating. Much of the world of ordinary everyday life is trivial and unimportant to the physicist as a physicist. The trees and cars that I can see from my study window may appeal to the aesthetic sense of a physicist, or be convenient ways of getting about in the world, just as they represent these things to all of us. But important as trees and cars as objects in the world may be in a professional context to economists or town planners, to the physicist they are arbitrary and uninteresting. The lesson to be learned is that the importance of things in our ordinary lives is no pointer to their scientific interest, and vice versa. Which is, perhaps, one of the reasons why science seems to be so strange and irrelevant to non-scientists. And until we have that grand unified science tied together with a grand unified theory of everything, the crucial divisions for one kind of scientist will often be irrelevant for another. Quarks and superstrings have no place, yet, in ecology.

With regard to these general aims, biology, including the social sciences, is no different from physics, chemistry or any other science. The most important job of the biologist or social scientist is to establish the smallest number of the most powerful principles that can be used to explain living phenomena, and to say what the biologically or socially important units of life are. These latter are the 'things' that we want to put our fingers on and measure, and are the causal mechanism of our general principles. And as with other sciences, there is no a priori reason to suppose that our ordinary experiences and divisions of the world will be congruent with those dictated by powerful scientific understanding. It is in this context that the search both for the most general formulation of evolutionary theory, and for how living things are to be partitioned into units that are significant for such a formulation because they are the physical embodiment of the processes of evolution, is to be understood. This carving of living systems at their evolutionary joints is what the phrase 'the units of selection' means in its widest sense. The units of selection are the embodiment, the mechanism, of whatever the general principles of evolution are.

We can now turn to what is known of the general principles that are the hallmark of any evolutionary process, and finally consider the issue of mechanisms.

THE GENERAL PRINCIPLES OF EVOLUTION

In a classic article on evolutionary theory published over twenty years ago, the American biologist R.C. Lewontin described Darwin's theory as comprising three principles. He called the first the principle of phenotypic variation. It states that different individuals in a population have different characteristics of structure and function, differences in both morphology and behaviour. The second, the principle of differential fitness, is that different phenotypes have different rates of survival and reproduction, depending upon the environment in which they live. Finally, the principle that fitness is heritable states that there is a correlation between the contribution of parents and offspring to future generations. That is, traits that contribute to the fitness of parents will most probably be inherited by their offspring. Any population in which these principles hold will undergo transformation through evolution. In the language of Chapter 2, evolution will occur in a population if the individuals in it vary in their characteristics; if their survival and reproduction are influenced by those characteristics; and if those characteristics are transmitted to their offspring.

All right, you say – a neat encapsulation of Darwinism, but not much new here. Well, not quite, because Lewontin suggested that there was a certain generality to the three principles. By this he meant that the mechanism of inheritance is not specified, only a correlation between parents and offspring is required. The inheritance could be genetic, cytoplasmic (that is, involving material in the cell that lies outside of the nucleus that contains the chromosomes) or cultural. What constitutes a population, and what are the individuals making up that population, need not be the same as is meant by the words 'population' and 'individual' in ordinary discourse. Also, the reasons for differential fitness are not stated and may involve many factors. Nor is there some single source of variation. In the standard case of biological evolution in sexually reproducing organisms, differences may arise from differences in genotype, or development, or both. However these three principles manifest themselves, and for whatever reasons – that is, whatever the mechanisms embodying the processes of generating variation, differential reproduction and heritability – when they are present, evolution will occur.

Now, what Lewontin made explicit about the generality of evolutionary principles had been implicit in the writings of evolutionary epistemologists from T.H. Huxley and William James onwards. Remember, evolutionary epistemologists are those who believe that the brain works like a 'Darwin machine'. That is, they believe that the transformation of brain and psychological states in time that occurs through the workings of the processes of learning and intelligence is the result of evolutionary processes operating *within* the brain. In other words, learning and intelligence are evolutionary processes embodied in brain mechanisms. This means that whatever those principles are that describe the evolutionary process operating between animals in the conventional sense of evolution must be the same as those that operate inside our heads. Donald Campbell, realizing the need for the spelling out within a simple scheme of the generality of evolutionary processes both *within* and *between* organisms, advocated, like Lewontin, a set of principles that he labelled collectively a 'blind-variation–selective-retention process'. Although there is some mismatch with Lewontin's three principles, the three essential ingredients of Campbell's overall process are similar to those of Lewontin. These are 'mechanisms for introducing variation' (Lewontin's phenotypic variation); a 'consistent selection process' (this is a combination of

Lewontin's differential fitness and its heritability); and 'mechanisms for preserving and/or propagating the selected variants' (Lewontin's principle that fitness be heritable).

There is a more economical, and for some purposes more convenient, way of stating Campbell's blind-variation–selective-retention scheme, which goes by the name of a 'g-t-r heuristic' (g-t-r standing for generate–test–regenerate). Heuristic ordinarily has the meaning of 'that which leads to discovery and invention', and is used here precisely because it indicates the inventive nature of selectional evolutionary processes. The g-t-r refers to three consecutive and continuous phases in the overall evolutionary process: the g phase is the generation of variants (the nature of which need not be specified, but which in the paradigmatic case would be genotypes, phenotypes or parts of phenotypes); the t phase is a test or selection phase (natural selection in the standard case); and the r phase refers to the regeneration of variants, combining previously selected variants and newly arising ones. Lewontin's variation (corresponding to the generate phase), differential fitness (corresponding to the test phase) and heritability of fitness (regenerate phase) are all present, if in somewhat different form. The g-t-r formulation is particularly useful when talking about learning and intelligence, and we will return to it often in Chapter 5.

None of these formulations makes absolutely explicit the presence of a transmission process by which selected variants are moved about in space and conserved over time. It is virtually stated both in Lewontin's principle of heritability of fitness and in the r phase of the g-t-r heuristic. The g-t-r heuristic repeats itself over and over and over in time, and for old variants to arise again in the r phase requires that they be conserved in time. So it is worth stressing that the propagation of variants requires some form of transmission system, that is, there must be some way by which the variants are maintained in time, and often this requires that they be moved about in space. Again, the mechanism of transmission need not be stated. Genetics is the paradigm transmission system for heritable variation, with genes being transmitted in space between organisms and by this means, in combination with a copying process, being conserved in time. But it is certainly not the only one.

Other kinds of general evolutionary principles have been put forward, for example those deriving from thermodynamic approaches to evolution. But the Lewontin–Campbell kind of approach is a distillation of neo-Darwinian principles and, for just that reason, much the most powerful and widely understood and accepted formulation. We can run these various schemes together to arrive at a set of processes comprising the generation of variants, the testing or selection of these, which results in differential fitness, and the transmission and regeneration of selected variants, their heritability, together with the injection of novel variants.

How does this abstract, mechanism-free formulation of evolution apply to Darwin in his several possible forms as fantasized earlier in this chapter? Consider again Darwin as psychologist. What he observed is the generation of diverse behaviours (behavioural variation); a selection process whereby some behavioural variants that lead to or are associated with 'good' events or lead away from 'bad' events increase in frequency (the testing phase resulting in differential fitness); and that the selected variants can then be re-membered (literally meaning that different parts are brought together again) or re-constructed (that is, rebuilt) at appropriate future points in time (the regeneration phase, which is also a form of temporal conservation and transmission process equivalent to heritability). As immunologist, Darwin was observing massive diversity in the membrane structure of lymphocytes (variation); he recorded clonal selection following the matching of an antigenic molecular structure to the antibody receptor structure on the membrane of an immature lymphocyte (selection); and he noted the retention in large numbers of 'memory' cells having the capacity to respond rapidly to the presence of specific antigen in the future, mixed in with the continuing generation of variant immature lymphocytes (regeneration, which again also serves temporal transmission). Darwin as a scientist of science was observing the near-random generation of ideas in what some historians of science call a pre-paradigmatic area of science (variation); the selection of a small subset of these as a result of the discovery of new phenomena or methods, or the production of a novel and persuasive theory (the test phase); and the propagation of these in linguistic or more formal mathematical form in books and journals, lectures, and so forth (this too is a spatial and temporal transmission system equivalent to heritability).

Of course, what Darwin the real man and naturalist did actually observe was the presence of large numbers of variant phenotypic forms (variation), some of which he inferred were better able to survive and reproduce than others (test phase). This was his great discovery, natural selection. Darwin knew that the fitter forms were able to pass on their characteristics to their offspring (heritability), and Mendel showed how they could be recombined in new forms (regeneration).

The point of all of this, of recounting what Darwin actually did and thought and the fantasies of what he might have done and thought had he been a different kind of scientist, is to show that however one arrives at an explanation of transformation in living systems by evolution acting between or within organisms, the processes that result in evolution are always the same. It is in the universal processes of variation, differential fitness and heritability, transmission of selected variants and their combination with new variants that we have 'universal Darwinism'. These are the processes that define a 'Darwin machine'. So the next question is, given these general principles of evolution, what is it in nature that we would like to put our fingers on and measure? What is it that provides the causal mechanisms for these principles? What are the significant units of evolution?

REPLICATORS AND INTERACTORS: THE CAUSAL MECHANISMS OF EVOLUTION

There are two ways by which one can move from the abstract, mechanism-independent set of principles of evolution to their mechanism-dependent embodiment. One uses more conventional ideas from within the evolutionary literature, while the other, referred to below as the replicator–interactor–lineage formulation, is a more recent arrival on the theoretical scene. It is likely that in the long run we will use both as circumstances dictate or allow, or as a way of buttressing one approach with the indications from the other as to what the mechanisms are. The first, the more conventional method, is a reversal of the procedure that we have used to establish these universal principles. That is, having abstracted the general processes by examining evolution in the form in which it is best understood, which is its conventional guise of that leading to adaptation and speciation as well as the functioning of the immune system, we reverse the direction of the enquiry and identify, in its various manifestations, the physical mechanisms that cause the transformations in time that we call evolutionary.

The results are obvious for the between-organism evolutionary forces that lead to adaptation and speciation, where the mechanisms are those already described in Chapter 2. Devices for generating variants exist at the genetical and developmental levels, which are tested by differential reproduction, and then the genetical bases of the more successful phenotypes are transmitted to offspring whose newly constituted genotype will add in additional and new sources of variation through mutation, chromosomal and genetic segregation, recombination and interaction. And so on, through endless cycles of generating variation, testing and then regenerating variation. The causal mechanisms upon which we wish to place our fingers, the units of selection, are genes, developmental pathways and the phenotypes upon which selection acts.

In the case of the immune system, the primary source of variation lies in the way antibody genes are inherited as fragments by the long succession of lymphocytes that are produced throughout life in tissues like bone marrow, together with mutations in the assembled genes coding for the hyper-variable membrane structures of antibodies. The test or selection phase occurs when an antigen fits a particular membrane structure on an antibody. Transmission results from clonal reproduction of the selected cells, some of which, as 'memory' cells, are the source of a long line of descendent cells of identical structure which are the basis of acquired immunity. And regeneration results from continuing production of antibodies of variable membrane structure. In the same way, if learning and intelligence are the result of the brain working as a 'Darwin machine', undergoing transformations in time that are due to internalized evolutionary processes, then it will be the job of neuroscientists to identify the means by which variant brain states are generated, selected and regenerated.

Lewontin, in that original paper, used something like this procedure to identify a number of possible units of selection. These included self-replicating molecules, cellular organelles (parts of cells), cells (lymphocytes, for instance), individual organisms, populations and species. All are potential units of selection under certain circumstances. Species have more recently also been put forward as units of selection by punctuationalists like S.J. Gould. One such punctuationist scenario goes like this. For reasons unknown, large numbers of species are produced within a relatively short period of time, geologically speaking. Some species are well suited to survival in the circumstances prevailing at that time and endure in time. Other species are poorly adapted to circumstances and become extinct. With the passage of time there comes another geological 'moment' in which some existing species continue, whereas others give rise to new species, and competition between them results in a new period of stasis in which some of the new species, and some of the old, continue to exist, whereas others become extinct. The scenario is rather short on actual mechanisms. Only species are there for us to 'touch', but the processes of variation (of species), selection (between species) and regeneration (a mix of new and old species) are all easily recognized. As Gould puts it, 'The same processes of variation and selection operate . . . but they work differently upon the varying materials.'

The second way of proceeding, the replicator–interactor–lineage (RIL) formulation, is based on the assumption that the processes of evolution are invariably accompanied by certain entities whose existence is proof of the evolutionary processes of which they are the constituent causal elements (in the case of replicators and interactors) or consequences (the lineages). So identifying the replicators, interactors and lineages not only proves that one is in the presence of evolutionary processes, but it also identifies the causal mechanisms by which those processes occur.

A simple description or definition of replicators, interactors and lineages would be too abstract an exercise to make much sense as an introduction to what they are. The best way of doing this is to tell the story of how these ideas came to be put forward at all, because it is only in this context that they make sense and have meaning.

The story of the RIL formulation begins in the 1950s, but to get the whole notion into perspective we have to take a small metaphysical diversion. Evolutionary theory has always carried with it, usually in a peripheral and unimportant way, some sense of 'what evolution is all about'. Why does evolution happen at all? people have asked. For the good of whom or what does evolution occur? This question has been asked partly, I suspect, because evolution was such a shocking revelation about humanity's place in nature. And one comforting way of answering it is that evolution has been 'for the good of us humans'. If *Homo sapiens* has evolved just as other creatures have, then might it still not be that evolution has occurred precisely in order to give rise to modern human beings? Put this way, evolution as a progression towards modern humans does leave us with some dignity and a special place in the scheme of things. Whatever the reason for this kind of thinking, this answer is certainly wrong − if, that is, the answer to a metaphysical question can ever be right or wrong. The pathways traced by evolutionary change over thousands of millions of years can best be described as contingent or accidental. There is no species in existence today whose place has been assured by evolutionary history, and that includes ourselves. We owe our existence to sheer good fortune.

None the less, the question of what evolution is for continues to be asked and has been asked for over a hundred years. In the last century the answer was mostly that evolution occurs 'for the individual', in the sense that it is the individual that seems central to the evolutionary processes and it is the individual that is to be measured. It is the individual that struggles for survival, and it is the individual that triumphs in the survival of the fittest. All adaptations were seen as benefiting the individual organisms that possess them. In this century there was a tendency, especially in the period from about 1930 to 1960, for biologists to replace the individual organism with the species; the adaptations of individual organisms were explained as 'being for the good of the species'. Fit individual organisms were a protection against species extinction. Although not explicitly stated, what seemed to underlie the switch from the individual organism to the species was the appearance in evolutionary thinking of the general Western cultural world-view that what is long-lasting is somehow superior to what is short-lived. And, by definition, species as entities are longer-lived than the individual organisms making them up.

In 1962 a Scottish biologist, V.C. Wynne-Edwards, wrote an interesting and influential book in which he attempted to explain the many instances of social behaviour that appear to reduce individual fitness but benefit the fitness of others, so-called 'altruistic' behaviour, in terms of 'group selection'. That is, such behaviour has evolved because it is 'for the good of the group'. Take the example of winter flocking by starlings. What winter flocking achieves, claimed Wynne-Edwards, is an assessment of flock numbers which, in the light of weather conditions and the general availability of resources, results in the birds adjusting their breeding in the coming season on the basis of their overall numbers and the conditions prevailing in the winter period. If their numbers are large and the winter is severe, then the birds, as a group, will reduce their breeding so that their numbers do not outstrip the resources in the coming spring. But if their numbers are few and the winter is mild, then their breeding will be stepped up because they will have available to them the resources necessary to support increased numbers of birds. All of this is obviously 'for the good of the group', not for the good of the individual birds. Wynne-Edwards did not suggest that the birds make calculations and arrive at conscious decisions after general deliberation and debate. Well-tuned stress-responses to crowding and cold weather could mediate the effect without individual decisions being taken by individual starlings. Wynne-Edwards offered many other examples, with accompanying ingenious explanations, of the behaviour of individual animals actually being adaptations to protect resources for the group as a whole and to adjust population numbers in terms of what is good for the group. These are called group-selectionist arguments, in contrast to the individual-selectionist arguments of people like Darwin.

The response to Wynne-Edwards's book from several authoritative evolutionists was that he simply could not be correct. Take the example of the starlings again. What would happen in time is that a mutant starling would arise that does not have the stress-mediated inhibition-of-breeding adaptation. Such a bird would, over time, produce more offspring than other members of the flock, and if the mutation for uninhibited breeding is heritable, then its offspring will also be uninhibited breeders. Very soon the population will be taken over by such 'selfish' starlings. All adaptations that are for the good of the group are vulnerable to competition from individuals in this way and could not last for any appreciable time.

This argument against the group-selectionist views of Wynne-Edwards is correct. Group selection is simply not a credible explanation for a form of social behaviour that is so widespread among so many different kinds of species. Individual animals belonging to many different kinds of vertebrate species display such altruistic behaviours. So too do members of many invertebrate species, especially the social insects like bees and ants. Perhaps, by some extraordinary chance, a group-selectionist explanation might hold for one or a few species – that is, in a very few cases, the selfish mutants never arise. But for thousands of species, certainly not. The chances of this having occurred across tens of thousands of species over tens of millions of years are so small that no scientist will accept it as the proper explanation of altruistic behaviour. But this left evolutionary biologists with a problem, one which Darwin himself had recognized in the puzzle of the evolution of sterile insect castes. Behaviours that reduce the chances of survival and reproduction of one animal but benefit those of another or others – in other words, behaviours that are contrary to the doctrine of individual selection and cannot be explained by group selection – must, none the less, be explained somehow.

Beginning with a series of technical papers first published in 1964 by W.D. Hamilton, an English biologist, and then enlarged upon by the American evolutionist G.C. Williams in a most elegant book in 1966 and by Richard Dawkins in *The Selfish Gene* in 1976, the answer that emerged is that such 'altruistic' behaviours, which cannot be for the good of the individual or the group, are in fact for the good of the gene. Consider the example of the 'broken-wing display' of many species of birds. A predator that comes too near a nest containing young birds is distracted by the parent rushing away from the nest and 'pretending' to be injured, thus luring the predator away from the nest but at considerable potential risk to itself. The way this is explained is that the bird giving the broken-wing display may put itself in danger, but by doing so saves its offspring. Parent birds, like almost all sexually reproducing vertebrates, share half of their genes with their offspring. If the parent were to be caught and killed by the predator as a result of the broken-wing display, the offspring, with many of the parent's genes, may survive because the predator had been distracted by the parent. Thus the parent's genes survive in the offspring. The altruistic behaviour is therefore explained by it being an adaptation for the good of the parent's genes, not for the good of the parent as an individual organism, nor, as a natural extension of the argument, for the good of the offspring as individual organisms.

This explanation of 'altruistic' behaviour in creatures like birds and bees is also almost certainly correct. And if *these* adaptations have evolved for the good of the genes, why should not *all* adaptations have done so? Why should not all evolution be for the good of the genes?

So what we see here is an explanation, a metaphysical account, of evolution in terms of what is at its centre. When metaphysics intrudes so directly into science it resonates oddly and uneasily with more traditional and stringent forms of explanation. The question of 'what evolution is for' is no exception. But it does serve to pick out what should stand at the explanatory core of evolutionary theory; some have argued for species, others for groups, individuals or genes. Although there is no single agreed answer as yet, by and large it is thought that it is either individuals or genes that evolution 'is for', and the most consistent and powerful argument within the context of conventional evolutionary theory, I believe, has been for the central role of the gene. If the gene selectionists are right, then, in this metaphysical sense, it is the gene that is the 'unit of selection'. As Dawkins has put it: 'all adaptations are for the preservation of DNA; DNA itself just *is*'. (DNA, short for deoxyribonucleic acid, is the name of the chemical substance of which genes are made.)

So, strange as this metaphysical approach may seem, it does lead to a specific notion of what the 'unit of selection' is for conventional evolutionary theory. Or at least it results in a more specific notion than comes from using the phrase 'unit of selection' to refer to a rather large aggregate of entities through which occur the processes of variance generation, selection, transmission and regeneration of variants as in the Lewontin–Campbell formulation. Further, Williams and Dawkins have sharpened up the notion even more by substituting, for the metaphysical idea of what evolution 'is for', a more concrete property, namely long-livedness.

Williams had a specific and good reason for concentrating on longevity as the defining feature of the unit of evolution within the context of conventional evolutionary theory, and it is important for the reader to understand the *conventionality* of Williams's, and other mainstream theorists', perspective. His view of evolution is one that is set within slow and gradual change being wrought over geological time to be measured in tens of thousands, hundreds of thousands or even millions of years. Williams was concerned to arrive at the most simple, lucid and powerful explanation of how adaptations can arise and be maintained *on this kind of time scale*. From this point of view, an adaptation is not a transient characteristic of a small number of organisms, but something that increases the chances of survival and reproduction in those creatures that manifest it, and that, being heritable, will spread throughout a breeding population and hence become a widespread and sustained characteristic of most of the individuals making up that population over long periods of time.

For example, we have no reason to believe other than that most polar bears have been white for some hundreds of thousands of years, or that most honey bees have been communicating the location of food sources to one another for a long, long time, or that the human ability for manual manipulation of objects with a high degree of delicate precision dates back to the beginnings of our species. These sustained characteristics may not be what evolution 'is for', but they are how evolution gets its results. Maintaining adaptations *for as long as they are needed*, which for conventional evolutionary theory is a very long time indeed, is the crucial 'objective' of evolutionary processes. Now, this sustaining role cannot be centred upon and carried out solely by the individual phenotypes bearing the adaptation, because, unlike adaptations, individual organisms are indeed transient and ephemeral things. So too, because of the ways genes are mixed and sorted by sexually reproducing organisms, are the genotypes of creatures like ourselves. *What do have the requisite constancy for sustaining adaptations across geological time are genes.* That is why, for Williams, genes are prime candidates for being the units of selection. Such units must have a 'high degree of permanence and a low rate of endogenous change'. Incidentally, to complete the circle, this is how the metaphysics creeps in. Genes are the entities that can do the job, and from this Williams and Dawkins arrive at the view that genes are therefore what the microevolutionary processes 'are for', namely the 'continuance of the dependent germ plasm' to use Williams's own words. It isn't a logical transition, but as I have said above, it certainly does concentrate the mind.

Now, nothing in biology is for ever. It has long been known that there is a continuous replacement of molecules throughout the structure of living things, including their genes, and any creature that is sufficiently long-lived, as we are, will have many complete turnovers of molecular structure during a lifetime. So what is sustained over time, and particularly in their passage from one organism to another during reproduction, is not the literal molecular configuration but the overall structural configuration of genes. It is the structural sequence that is sustained, because it is the sequence that codes for the information that results in the development of adaptations in the phenotype.

Dawkins knew that DNA itself, here on Earth, is not for ever. He also recognized that extraterrestrial life forms are unlikely to have DNA as the chemical code by which information is encoded and transmitted across generations. But, of course, they must have some other chemical playing that role. For both of these reasons he suggested that universal Darwinism, that set of evolutionary processes that is the same anywhere in the universe, requires the concept of a key player that he called the 'replicator'. One of the principal qualities of this entity is that it can make copies of itself. Any life-form-sustaining adaptive structure requires the transmission of information coding for that structure between individuals, but in such a way that no or few errors creep in, or with built-in devices that can correct for such errors. A copying process of high accuracy is therefore required, because, with faithful copying, information is indeed for ever. Dawkins coined a fine description of the trinity of attributes that a replicator must have: longevity, fidelity and fecundity. Replicators must be long-lasting, capable of making good copies of themselves and fertile in making as many good and long-lasting copies of themselves as possible. Genes, of course, are the paradigmatic replicators. But they are not the only ones, even here on Earth.

Adding another theoretical entity, the vehicle, to the scheme allows the Williams–Dawkins formulation to dovetail into one of the oldest and most important distinctions in all of biology: that between the genotype and the phenotype. Remember, the genotype is the total of all the genetic information within an individual, whereas the phenotype is the expression of that information in the flesh-and-blood individual that develops via a series of highly complex interactions with the environment. You and I are phenotypes; the genetic information contained in the fertilized egg at conception from whence we came, and which is still present in almost every cell in our bodies, is our genotype. Now, selection operates at many levels. Natural selection traditionally is conceived as selection operating *on* the whole phenotype, while it is the gene that is selected *for*. Exploiting this distinction, Dawkins advocated that phenotypes should be seen as devices that genes, the replicators, ride about in; and, in accordance with the abstract nature of universal Darwinism, he suggested the term 'vehicle' for that in which replicators shelter and ride, and upon which selection acts. The vehicles are the instruments of replicator preservation. More accurately, because of selection, the vehicles are the instruments of differential replicator preservation.

Dawkins's formulation of replicators and vehicles has aroused considerable interest within both the scientific and the philosophical communities of biologists. One of the characteristics of the way Dawkins has developed the ideas is that it casts the replicators into the role of controllers or manipulators of passive vehicles, whose role is to provide shelter and transport for the replicators and upon whose hapless heads the blows of selection fall. Now, back in the 1960s, Waddington had argued something like the reverse case. Evolution, Waddington had asserted, requires both a stable memory store (analogous in role to replicators) and what he called an operator component. The operator is equivalent to a vehicle, but Waddington made it anything but a passive entity. He invested it with dynamic, reactive qualities, one of which is the capability in some cases of altering its own environment. We will return to some of these characteristics in subsequent chapters. The operators are what make life 'interesting', according to Waddington, whereas the memory stores are rather dull and unreactive – DNA is, in fact, an exceptionally inert chemical. So as you can see, the spirit of Waddington's depiction of evolution is somewhat opposed to that of Dawkins.

More recently, David Hull, an American philosopher of biology, has resurrected the 'operator' type of idea and called it an 'inter-actor'. Interactors are entities that, according to Hull, can be defined as interacting as cohesive wholes with the environment to result in differential replicator selection. Phenotypes, as might be expected, are paradigm interactors, but Hull's argument is subtle and complex and under certain circumstances he includes genes, as well as chromosomes and sex cells, as interactors. But the most important feature of the interactor, and what contrasts most strongly with Dawkins's vehicles, is the connotation of them being causal agents in their own right: interactors do things that are not reducible to the orders served up by the replicators riding about within them. This too is an issue that we will return to in the next and later chapters. All that we need register here is that the notion of an interactor *is* different in this respect from that of a vehicle – and, in my view, it is to be preferred precisely for that reason. If interactors did not contribute in some way independently of replicators, then all of evolutionary theory could be reduced to replicator terms. Evolutionary theory cannot be so reduced, precisely because, to persist with the metaphor, interactors are not always blindly obedient vehicles wholly manipulated by replicators.

The microevolutionary interplay between replicators and interactors has macroevolutionary consequences. 'Lineages' are defined by Hull as entities that can change indefinitely through time as a result of replication and interaction. Neither genes nor organisms can *be* lineages, because neither can change very much, and certainly not indefinitely. But both do enter into lineages, as is the case with a population that can indeed change greatly over time. Species, of course, are the entities that traditionally make up the lineages of conventional evolutionary theory.

This, then, is the RIL formulation. In summary, if entities that can make copies of themselves (replicators) are propagated in space and conserved in time because of the differential extinction and proliferation of interactors, these will in turn lead to historical changes in lineages, and evolution will have occurred. I have presented it as simply as possible. In the hands of Dawkins, Hull and others it is an altogether more complex and subtle set of ideas. Partly this is because living things do not always conform to the patterns of sexually reproducing animals, in terms of which we are so used to thinking. For example, in asexually reproducing organisms, large parts of the whole phenotype may be replicators; certain plants are notoriously difficult to fit into the kinds of divisions that animal biologists are used to; and even in sexually reproducing organisms, what is a replicator may vary with the degree of genetic heterogeneity of the breeding population and with gene linkage. There are also more than just one interactor, and it may be that sometimes the same entity can be simultaneously replicator and interactor. One must keep this kind of complexity in mind, because while the RIL formulation may have a spare elegance about it, it can also be rich and complex.

Yet even in its most simplified form the RIL formulation may seem unduly remote, abstract and fussy. But if evolutionary theory is to come into its own as a really powerful explanation of the transformation of living systems in time, then it may well have to abandon its common-sense terms and their application in a conventional way to 'ordinary' evolutionary events. To repeat the point made earlier in this chapter, the biologically significant ways of dividing up the living world may not agree at all with common-sense or ordinary experience.

How well, if at all, does the RIL formulation apply to that other system that we know is transformed by evolutionary processes, but which occurs within organisms, namely the immune system? Here the hyper-variable regions of the antibodies, which can be identically copied by clonal reproduction, are the replicators. It is the immature lymphocyte upon which selection acts, and since lymphocytes are coherent wholes and their differential reproduction determines differential replicator survival, it is the lymphocyte that must be the interactor. The lineages are the many lines of descendent memory cells that bear the history of the antibody–antigen interaction of the individual during its lifetime.

The RIL formulation, then, does seem to work with the immune system, though immunologists might wish to substitute other components within the immune system for those that I have picked out here as replicators, interactors and lineages. What of the punctuationist approach to species selection? Here I am less at ease. The replicator must, I think, be the gene pool (that is, the total genetic constitution of all the individual members of that species), for that is the only possible entity that might make copies of itself. Selection operates upon the totality of phenotypes making up the species, and so that must be the interactor. The lineage will be what taxonomists call the clade, which, roughly, is a higher-order group of related species.

Some will argue that the RIL concept is a less certain route to take in identifying the causal mechanisms of any evolutionary process, because, while it has proved itself well enough with evolutionary theory as conventionally applied to the formation of adaptations and species, it is much more difficult to apply to other forms of transformation in living systems that are attributable to evolutionary processes. There are at least three possible reasons for this difficulty. One is that the RIL formulation is an entirely correct way of proceeding, but evolutionary theory only really applies to this conventional formulation where the processes act between organisms. I reject this argument completely. If nothing else, the immune system is one that works by way of the processes of generate, test and regenerate – it is, in other words, an internalized evolutionary system. If the immune system is a 'Darwin machine', why not other internalized systems whose essential functions require them to be transformed in time? Another possibility is that the way the RIL formulation puts flesh on the bones of the evolutionary processes is extremely limited and has no general use at all beyond conventional theory. I think this unlikely, but it cannot be dismissed out of hand. The third reason may be that, while we may not be able to use the RIL formulation successfully now in all instances of evolution, both between and within organisms, this is because we do not yet know enough about these other forms of evolutionary change to be able to do so. This, I suspect, is the real reason.

SUMMARY AND CONCLUSION

I have tried to present in as simplified a way as possible some complicated ideas, and it might be useful, therefore, to provide a résumé of the argument.

The history of evolutionary epistemology and the fantasy of Darwin in the role of different kinds of scientist were used as a way of introducing Dawkins's notion of universal Darwinism. We then considered that every area of science has to establish the fundamental laws that govern the way things work in that bit of the universe that is covered by that science; and that there are certain basic entities, the causal mechanisms, that determine the nature of those laws. The job of the scientist is to formulate the laws and discover the causal mechanisms. In biology, one form of transformation of living creatures, or parts of those creatures, in time, is evolution. What evolutionary biologists must do, therefore, is determine the essential processes by which evolution occurs, whether it occurs between organisms (which is evolution as conventionally understood) or within organisms. These processes are universal. Then, having established these universal processes, they must identify which parts of living systems are those basic mechanisms that cause these processes to occur, that is, which constitute the units of selection. These, of course, will vary depending upon whether the processes of evolution are occurring between or within organisms, and in the case of the latter, which organ system is involved. Finally this search for mechanisms, for the units of selection, can be undertaken in one of two ways. One can either search for the generators of variation, the selection devices, and how the selected entities are transmitted and propagated; or one can seek to establish which are the replicators, which the interactors and which the lineages.

There remains the question of which is the better approach to identifying the units of selection: should it be a search for the mechanisms subserving the g-t-r (what I have labelled the Lewontin–Campbell approach), or should we be looking for replicators, interactors and lineages, the Dawkins–Hull approach? At present, we just do not know which is the best way to go about dissecting out the mechanisms, the units of evolution. With further refinement we might find that the two approaches eventually blend into one conception. That, for what it is worth, is where my money is. Further, the way that evolutionary processes actually do occur in the immune system, in the brain or at the level of the species is still largely unknown. But there is no reason not to expect considerable differences of detail. *The actual mechanisms in each case*, of course – and one cannot repeat this point often enough – *are entirely different*. Given these great gaps in our knowledge, we cannot yet be prescriptive but have to try both approaches, and probably change each, before we know with any certainty what is the best conceptual road to follow.

Finally, keep in mind the idea of a hierarchically structured evolutionary theory. A true hierarchical theory of evolution would not be based upon the relationship to each other of components of the *same* unit of selection or of the components of some *single* RIL formulation, which is the case for genotypes and phenotypes in conventional evolutionary theory. Instead it must be based upon a hierarchy of *different* units of selection or RIL formulations. The hierarchy will comprise different levels at *each* of which there are identifiable processes of variant generation, selection and regeneration. Such a hierarchy might equally well be depicted as having levels, each of which contains its own replicators, interactors and lineages. And it is precisely this kind of hierarchical conception that we will build over the next two chapters in order to develop a theory of knowledge.

Now that we have the conceptual tools necessary for building such a theory of knowledge, we can turn to the subject of behaviour.

SUGGESTED READING

Universal Darwinism, a recently coined phrase though an old idea, is treated nowhere, apart from this chapter, in a simple and accessible way. The following are suggested for readers who want to pursue these matters and are prepared to wade through some highly technical writings.

Campbell, Donald T. (1974) 'Evolutionary epistemology', in P.A. Schilpp (ed.), *The Philosophy of Karl Popper*, Volume 1, pp. 413–63, La Salle, Ill.: Open Court. Darwinism applied to learning, thought and science by the founding father of modern evolutionary epistemology.

Dawkins, Richard (1983) 'Universal Darwinism', in D.S. Bendall (ed.), *Evolution from Molecules to Man*, pp. 403–25, Cambridge: Cambridge University Press. The argument for evolution anywhere in the universe occurring by Darwinian processes.

Edelman, Gerald M. (1987) *Neural Darwinism*, New York: Basic Books. A complicated and technical account of the brain as a 'Darwinian machine'.

Behaviour without Thought

We human beings, in common with most other animals, move about in the world. The reason for this is that we are not autotrophs. Autotrophs are organisms, like plants, that can manufacture the complex compounds of life, including the substances that fuel living processes, from simple inorganic ingredients. This in turn means that autotrophs are able to survive in exclusively inorganic surroundings and hence they can stay in one place and manufacture the essential ingredients of life simply by utilizing the materials that wash about them. Water and certain simple elements such as nitrogen, magnesium and phosphates are extracted from the soil, and carbon dioxide is taken from the air. From these are constructed the basic building blocks of life, and, most importantly, they are also used in plants for the transformation of light energy from the sun into chemical energy, a process known as photosynthesis.

Photosynthesis, though probably not the original means by which the energy of the sun was channelled into living things, is the fundamental process that fuels life on Earth now, and it has been so for some thousands of millions of years. Plants can do it, but non-autotrophic organisms like ourselves and other animals cannot. The way that animals obtain their fuel is by eating plants, or by eating other animals that eat plants, or by eating animals that eat other animals, or some combination of these. Whatever the particular 'life-style' might be that enables animals to break into the chain that begins when solar energy is trapped into a molecular web by photosynthesizing plants, the consequences for animals of having to do so are momentous. This is because animals and plants, or the products of plants like their fruit, are not evenly distributed in the way that light and air are evenly distributed. A plant or an animal is a container of chemical energy that is localized in a particular region of space and, in the case of a potentially moving animal, one that may change its position in space. Even when plants or animals aggregate they are almost always unevenly distributed in space. These discrete packages of chemical energy, therefore, form a heterogeneous patchwork of energy resources for the animals that have to contact and exploit them if they are to stay alive. The only way that animals can do this is by moving themselves about in space.

This requirement for movement, and the associated consequences of that movement, have been fundamental selection forces in the evolution of some kind of organ for the co-ordination of movement (the brain and nervous system), for the evolution of organs that produce the movements and for the evolution of sense organs that allow for better-directed movement. As we shall see, it has also been the fundamental selection force in the evolution of learning and intelligence.

The origins of behaviour in animals, therefore, lie in this general need to move about in order to exploit scattered energy resources. Subsequently, of course, this capacity for movement became enlisted in the service of a large number of other functions, such as escaping from regions of danger, contacting mates and thermo-regulation, to name just a few. Yet the ordinary meaning of the term 'behaviour' is even wider; its root is the Latin verb *habere* (to have), so that its common meaning extends to 'having possession of' and 'being characterized by'. Now, behavioural biologists simply never use the word in these senses. Behaviour refers to *doing*, not having, but even then the range of meaning is bewildering unless one has some general way of tying all instances of behaviour to some broad and unifying conception.

To see how this can be done, consider a few examples. A bird that flies from one place to another because it is foraging or escaping from danger is indubitably behaving, the doing involving a shift in space of the whole creature. That same bird is also behaving when it manipulates an object like a seed or an insect with its beak, though the bird in its entirety may be stationary at the time. In this case the doing is limited to just a part of the animal, that part being what anatomists would refer to as a specialized effector organ, which is a part of the body that has evolved a form specific to particular kinds of doing. So the beaks of birds of different species may have different shapes, depending upon whether the individual birds use their beaks primarily for cracking seeds, piercing fruit or shovelling up prey, the nature of the doing itself having become a part of the selection pressures that shape the forms of animals.

Now, that same bird is also behaving when it sings, even though such behaviour is far removed from movement of either the whole animal or a part of it like a limb. Certainly it involves the action of a specialized organ, the syrinx, which is the avian equivalent of the larynx, but what makes it behaviour? Well, the answer is the same thing that makes pecking at a seed behaviour or flying from one tree to another also an instance of behaviour. It is an act of doing *in order to have an effect upon the world* – in the case of singing, the attracting of a female or the warning off of neighbouring or passing males.

The point can be made clearer with other examples. When a mature male dog lifts its leg and urinates on a post or tree, it is behaving. Female and immature male dogs do not lift their legs, and so when they urinate they are not behaving. What is the difference? It is not the leg-lifting as such. The difference lies in the mature male dog doing something when urinating that other dogs do not do, and this act of doing is the scent-marking of its territory at a nose height just right for other dogs to detect. That is why the leg is raised. In this case urination is an act of *doing* that has the effect on the world of informing other dogs of its presence. Immature male dogs and female dogs do not scent-mark, and so they do not lift their legs when urinating, the latter simply being a manifestation of renal physiology and nothing more.

Coughing in humans is another example. If I cough because I have a cold or have inhaled some crumbs or noxious smoke, then that coughing is a part of normal respiratory function – the action of coughing serves to expel foreign material from the lungs. But if I cough because I want to gain attention, or sympathy, or both, then that instance of coughing is behaviour because it is an act of *doing* that serves to change the actions or attitudes of others towards me: I am coughing for a reason that is not intrinsic to lung function but is in order to effect some specific change in my world. A crucial part of the distinction being drawn is between activities that are directed towards effecting change in the world outside of the behaving creature or in altering the relationship between the creature and some aspect of the outside world, and activities that are accompaniments to, or consequences of, normal internal bodily function.

Jean Piaget captured and added this sense of how to demarcate behaviour from other actions or changes of state of animals by defining behaviour as 'all action directed by organisms towards the outside world in order to change conditions therein or to change their own situation in relation to these surroundings'. Thus behaviour for Piaget was 'teleological action', that is, goal-directed or end-directed adaptive action.

I am going to use Piaget's definition throughout this book, so we had better be clear as to what kind of work one expects a definition to do. A definition is not something that is necessarily true, and it does not come to one by some form of revelation. All that a definition does is help to explain what one means, and it does this by leaping out of the page at the reader rather than having that meaning surreptitiously attached to a word without any form of announcement. So when one writes 'By behaviour I mean teleological action' there cannot be much doubt as to what one does indeed mean. Secondly, definitions serve a sort of propaganda purpose in that they are an occasion for making clear one's theoretical predilection. I believe that the concept of adaptation is so crucial to biological science that it is worth wrapping all of behaviour up within it. Piaget gives us the excuse for doing this and allows a direct connection to be made between the notion of behaviour and everything stated in Chapter 2 about the characteristics of adaptations.

In defining behaviour in adaptational terms, however, the reader must not be mistaken into believing that all actions or activities are adaptive. Clearly they are not. As I type these words, my chest is moving up and down to the rhythm of my breathing, but that is just another manifestation of respiratory function. And my shoulders are swaying about because of the movement of my fingers. The finger movements are behaviour because they match the configuration of the keyboard in the manner typical of the relational characteristics of all adaptations, whereas the shoulder movements are simply a mechanical consequence of my fingers being connected to my arms which jiggle about and cause shoulder movements. Many instances of action and activity lack either the matching characteristic of adaptations or their fitness-enhancing function. They may be of great interest to a physiologist, but they are just not important to a behavioural biologist.

Defining behaviour in a way that links it to adaptation, incidentally, allows one to distinguish also between inaction or inactivity that is behaviour and that which is not. An animal that is still because it is sleeping is not behaving (unless the theory that sleep is an anti-predatory device is correct), whereas one that is frozen into inactivity in the presence of danger is behaving. The distinction turns not on the presence or absence of consciousness or awareness, but on whether the inaction has been selected for its adaptive significance or is merely a consequence or correlate of some other bodily function. In the case of sleep, the inaction is like coughing when one has a cold – it is a part of some other function. But freezing to avoid detection or attack is an adaptive inactivity by any definition, and so qualifies as behaviour.

We will return in the next section to the consequences of thinking of behaviour as always being adaptive; first we must consider the causes of behaviour.

Scientific accounts are cast in terms of causes. Scientific theories are causal explanations usually dressed up in the language of the processes and mechanisms discussed in the previous chapter. A science of behaviour, therefore, must look to its causal explanations if it is to be any kind of science at all. So what kind of answer can be conjured up when one asks about the causes of behaviour? In order to be specific, let us use the example of that flying bird. The example is a convenient one because Ernst Mayr, the evolutionist, published a famous paper in 1961 on the concept of causation in the biological sciences, his analysis centring upon the migratory behaviour of birds. What are the causes of this migratory flying behaviour? I am going to rule out one kind of explanation at once, because I do not believe that it is a causal one. This is the 'wings flapping' and 'hollow bones' kind of argument. Birds do not fly *because* they have feathers, wings and bones structured to have as little weight but as much strength as possible. All of these are adaptations *for* flight in that they are enabling features without which flight would either not occur at all or be much reduced in efficiency. But they cannot in any sense be claimed to cause flight. What causes flight, specifically migratory flight, can be divided into two sets of causes. On the one hand there are what Mayr called the proximate causes. In a migratory species of bird one of these proximate causes might be what is known as photoperiodicity, which is a sensitivity of the animal to the length of daylight in a twenty-four-hour period: when the total drops below some value, it triggers the behaviour of migration. Another form of proximate cause signalling a change in the seasons might be the temperature on the day that the bird begins its migration. In both cases the proximate cause begins with a change in the environment that is detected by the bird. This information is conveyed via sensory pathways to the central nervous system of the animal, leading to the activation of those parts of the brain that control migratory flight, and the appropriate behaviour then occurs.

There are a number of ways of characterizing these seemingly different kinds of cause. One is in terms of the sorts of scientist who investigate them; in this case laboratory-based or 'bench' biologists, like physiologists. Another is in terms of the kind of question asked by bench biologists, which is 'how?'. And when 'how?' questions are raised with regard to behaviour, one of the main channels of investigation is to uncover mechanisms in the brain of the animal – a kind of 'what does what?' in the brain.

The second set of causes were labelled ultimate by Mayr. Let us assume that our migrating bird is an insectivore. In that case, one of the ultimate causes of migration is that the bird is the product of a long history of selection caused by the decline in the insect population during the autumn and winter months, one adaptive response to this decline being to migrate at that time of year to warmer climes where the insect population is more numerous. Another way of saying this is that the history of selection has led to that bird having a particular genetic constitution and it is that constitution that is the cause of migration. Such a genetic cause, of course, is mediated by the genes finding expression in the development of a brain so structured that in the autumn the bird migrates. Like the study of proximate causes, ultimate causes can be picked out by certain characteristics. Scientists with an interest in ultimate causes tend to work with animals under more natural conditions than the laboratory, often 'in the wild', so that the adaptive value of the behaviour can be assessed. They emphasize 'why?' questions designed to establish adaptive function. And as I have just indicated, the thinking about cause tends to focus on genes rather than on brains.

From what was said in Chapter 2 it will be realized that the phrase 'evolutionary cause' can be substituted for 'ultimate cause', tapping as it does into just that set of features that characterize evolutionary thinking. Proximate causes, on the other hand, could just as well be called physiological or functional causes. The heart of the distinction between them is best summed up by Mayr himself. Ultimate causes concern the *encoding* of information into the DNA structure of genes, whereas proximate causes deal with the *decoding* of that information by developmental processes resulting in an organism with specific structural and functional features. This is indeed a useful distinction. None the less, it is worth pointing out that ultimate and proximate causes are not in any deep or fundamental way different from, or independent of, one another. What is encoded in genes depends at least in part on the competence of the decoded product – this is the eternal linkage between genes and selection acting upon the organism. This means that every attribute of an organism, including its behaviour if it is an animal, has *both* ultimate *and* proximate causes attaching to it. In the case of behaviour, it is on the one hand always and simultaneously the result of brain-developmental processes occurring in a particular environment and the subsequent functioning of the brain; and on the other the result of the adaptive value of the behaviour in the animal's forebears and the manner in which it contributed to their fitness, hence determining the transmission of the genes coding for that behaviour and their eventual inheritance by the animal in question.

It is nothing more than a matter of taste as to which of these kinds of causal explanations one ends up working on as a scientist, and which one emphasizes when explaining a phenomenon in biology. This book is an account of knowledge cast largely in terms of ultimate causes, but this does not mean that we are going to be ignoring proximate causes entirely. Far from it. Some conceptions about how the brain works to generate a particular kind of knowledge require some attendance to proximate brain function, but that is for a later chapter.

ADAPTIVE BEHAVIOUR

Defining behaviour as adaptive action or inaction does not relieve us of the obligation to demonstrate if possible that many actions are indeed adaptive. However, as discussed in Chapter 2, despite the potency and great consequence for biology of the concept of adaptation, adaptive attributes are disconcertingly difficult to identify with certainty and do not easily lend themselves to measurement. In large part one has to rely on the extent to which an attribute satisfies at least one, and if possible more than one, of the criteria that qualify it as an adaptation. The best way of doing this with behaviour is to consider some examples from the ethological literature.

Ethologists are behavioural biologists whose primary interest is in studying behaviour in the context in which it evolved, which is to say 'in the wild'. Ethology has its roots in the study of behaviour by the nineteenth-century naturalists, but only graduated to the status of a distinctive scientific discipline with the work of the German biologist Konrad Lorenz in the 1930s. One of Lorenz's earliest collaborators was a Dutch biologist, Niko Tinbergen, and Lorenz and Tinbergen together are generally held to be the founders of modern ethology. Ethology has changed a great deal in recent decades, and to distinguish between present-day ethology and that of its founding fathers I will refer to the latter as classical ethology.

Central to classical ethology was the comparative method, itself one of the great contributions of nineteenth-century biology. It compares the characteristics of closely related species, and accounts for differences between them in terms of adaptations to differing environmental requirements. The chief idea of the method is that closely related species will be largely the same because they have only recently diverged from common ancestral stock, and when differences are present it will be mainly because of the effects of different histories resulting from the effects of natural selection in differing environments. This is known in the trade as divergent evolution. The classical ethologists were great exponents and practitioners of the comparative method as applied to behaviour.

The sea-gulls were one group of closely related species extensively studied by Tinbergen and his students. Most species of gull, including the well-known herring gull, are ground-nesting birds. The striking exception is the kittiwake, which is a gull that nests on tiny ledges of very steep, often sheer, cliffs. (In fact, there are a couple of species of gull that nest on not-very-steep cliffs, and their behaviour is roughly intermediate between that of kittiwakes and ground-nesting gulls.) Now, the kittiwake evolved from an ancestral ground-nesting gull; the move to nesting on steep cliff faces was driven by the benefits such a nesting site provides in terms of reducing the predation from the likes of foxes and crows that plague the lives of ground-nesters. But the steep cliffs also impose particular selection pressures upon the kittiwake that are different from those imposed on its ground-nesting cousins. One might therefore expect behavioural differences to be present between cliff-nesters and ground-nesters that reflect these differences, and that is exactly what the ethologists found.

There are over thirty well-documented differences between the behaviours of kittiwakes and those of ground-nesting gulls. For instance, ground-nesting gulls remove from their nests materials that would make them conspicuous to overflying predators. Kittiwakes, on the other hand, can best be described as messy nesters. They do not clean their nests, but then because they nest on ledges set into steep cliff faces, their chicks are not preyed upon by the carrion crows and herring gulls that take the young of ground-nesters. (Herring gulls have two roles to play in this story. Like most sea-gulls they are ground-nesters and hence subject to similar pressures of predation to other ground-nesting species. They are also, however, predators themselves, preying on the nests of other gulls.) The nests of kittiwakes are also differently constructed from those of ground-nesting birds, not only in terms of the methods and materials of construction but also in form, the cup of kittiwake nests being deeper than that of ground-nesting gulls. Hence the likelihood of losing an egg from the nest – which for kittiwakes is usually an irretrievable event, since the egg will tumble down the cliffs – is much less for kittiwakes than for ground-nesters, for whom egg retrieval is a relatively easy task. Differences in aggressive behaviours, appeasement gestures, mating between adults, foraging for food and the feeding of chicks have also been well documented.

Consider one other specific example. When a ground-nesting bird like the herring gull threatens to attack another herring gull, the bird that is threatened can appease the attacker by increasing the distance between itself and the bird that is threatening it. An uncommon additional appeasement gesture in adult ground-nesting gulls may also occur. This is known as head-flagging, whereby the threatened bird turns its head to one side and partly hides its bill in its feathers – as if to say, 'Look, I'm no threat to you. I don't even have a bill with which to peck back at you.' Head-flagging never occurs in young ground-nesters. With a kittiwake, because it lives on narrow cliff ledges, it cannot easily put distance between itself and another threatening kittiwake. And kittiwake chicks certainly cannot do so, since they cannot yet fly. How, then, do kittiwakes appease potential attackers? They do so through very pronounced head-flagging. The head is rotated through almost 180°, the feathers on the back of the bird's neck are raised, and the bill is completely hidden. Even kittiwake chicks head-flag in this exaggerated manner. So, what in ground-nesting gulls is an uncommon, minor, rather half-hearted appeasement gesture has evolved into the most prominent and common form of appease-ment behaviour that can be observed in kittiwakes, including the chicks.

The behavioural differences recorded between ground- and cliff-nesting sea-gulls appear to fit the differences in their physical environments. The results of such divergent evolution are different behavioural adaptations, the claim that these differences are indeed adaptations resting on the criterion of the goodness-of-fit between the two components of the putative adaptation – that is, between the feature of organismic organization (in this case, a behaviour such as nest construction) and the matching order of some aspect of the world (cliff face or level ground in the case of the gulls). By this criterion of design, nest construction in the kittiwake is indeed an adaptation to cliff-dwelling.

Another sign of the adaptive nature of an attribute is when it can be shown that it contributes to the overall efficiency with which its possessor converts its own energy resources into offspring. I have already mentioned that, in comparison with the unsavoury nest state of kittiwakes, ground-nesting gulls are paragons of cleanliness. In a long series of studies carried out in the 1950s, Tinbergen and his students were able to show that the eggshell-removal behaviour of black-headed gulls is indeed adap-tive in terms of this measure of production and survival of offspring. They were able to demonstrate that the presence of broken eggshells in the nest attracts the attention of carrion crows and herring gulls, both of which are species that prey on the nesting grounds of black-headed gulls. They were also able to show just what it is about eggshells that attracts the attention of these predators. Now, from the evidence it is reasonable to make the assumption that gulls that do not remove the shells of hatched eggs from their nests are going to lose more chicks to overflying predators than are those that do keep their nests clean. The number of offspring that an animal produces is the classical measure of fitness. The traits that contribute to this measure are adaptations whose existence is owed to their selection in the past precisely because of this contribution to fitness. Nest-cleaning in gulls, therefore, can be inferred to be adaptive activity on this count, and hence to be behaviour by Piaget's definition.

A third sign of adaptive organization is complexity of structure whose occurrence by chance is too remote a possibility for one to be able to accept that happenstance is the proper explanation for its presence. Complex organismic organization is diagnostic of the action of concerted selection processes over very long periods of time. Are there, in terms of complexity, behavioural equivalents of the eye? High on the list of most ethologists in this regard would be the extraordinary behaviour of social insects such as ants and bees. This was certainly the case for Darwin, who in the chapter devoted to 'instincts' in *The Origin of Species* chose as examples for discussion slave-making in ants and cell construction by honey bees.

The cluster of behaviours that go to form the social activities of honey bees in particular is indeed extraordinary. It includes the fissioning of the hive; swarming and migration to a new hive site; the construction of the hive including cells, in Darwin's words, 'of the proper shape to hold the greatest possible amount of honey with the least possible consumption of precious wax in their construction'; the provisioning with food of both queen and larvae; the cleaning of the hive including the removal of dead larvae; and, most extraordinary of all, the famous bee dance by which foragers communicate the position and richness of food sources in a manner which includes a correction factor for the movement of the sun, which is their principal source of navigation.

All of these are components that add up to the success of the honey bee as a social creature. But since it is not the object of this book to provide a descriptive account of the wonders of animal behaviour, I must refer the reader either to the many beautiful accounts of social insects that are to be found in the widely available general texts on animal behaviour or in the more specialized writings of the likes of Karl von Frisch, Eugene Marais and E.O. Wilson, who do a far better job than I ever could in this regard.

The point that I am making here is simple and obvious. The

behaviour of an ant that goes out into the world, captures an ant of another species, returns home with its captive and sets it to work to serve most of the captor's needs cannot be viewed as a chance occurrence alone. Selection had to be acting to sculpt the component chance bits into a coherent, complex whole. The outcome is what Julian Huxley called the 'bundle of adaptations' that make these animals the astonishing and beautiful things that they are.

One of the inevitable consequences of accepting Piaget's definition of behaviour is that it follows that behaviour is a product of the evolutionary processes described in Chapters 2 and 3. From this in turn follow certain other lines of evidence regarding behaviours and their adaptive nature. One is that there must be, or must have been, variations in the behaviours that animals now have (evidence for the g phase or principle of variation described in Chapter 3); that selection will have acted upon these variations to result in better adapted behavioural forms (the t phase or principle of differential fitness); and the selected behaviours will be propagated across generations by some form of transmission system which may also give rise to further variation (the r phase or the principle that fitness is heritable).

Little evidence has been gathered on variation, largely because the attentions of behavioural biologists have been elsewhere, and because, I suspect, doing such work is viewed as being intrinsically dull. I have no doubt, though, that anyone looking for variation in, say, nest-building behaviours in birds will find it. The existence of transmission processes is much better documented. Behavioural genetics is a thriving discipline in its own right, and literally hundreds of behaviours, from simple tendencies (tropisms in the jargon of biology) such as a tendency to move away from gravity (negative geotropism) in certain species of insect, all the way through to the ability to learn certain things in mice, have been shown to be in part caused by genes. This does *not*, of course, mean that the behaviours in question are *only* caused by genes; and it does not mean that the nature of that causation is of a simple one-gene-to-one-behaviour form. But these behaviours demonstrate what the Piagetian position demands: if they are adaptive, then they must be products of evolutionary processes, one of which must subserve the maintenance in populations across successive individuals of the behaviours that have evolved. At least one transmission mechanism, and that the traditional genetical one, has been demonstrated by the behavioural geneticists to be in place, and so presumably has been used in the evolution of these behaviours.

There is yet another consequence of the evolutionary origins of behaviour, which is that through the action of divergent evolution one should see behaviour appearing in the form of a variation on a theme, or set of themes, in related species. The differences in the behaviour of gulls that were described earlier are one such example. Another is to be found in the very curious behaviour of male empeid flies, first described by the American behavioural biologist T.C. Schneirla. Empeid flies are predatory insects with powerful grasping jaws and poisonous salivary secretions. The females are especially fearsome and will often eat male empeid flies when they can get close enough to them. This is awkward for the male flies, which do indeed have to get close to the females when mating with them. The grim prospect that male empeid flies have to contemplate is that of being eaten by a female empeid fly during or immediately after mating.

This unpleasant disposition on the part of the females has been the selection pressure for the evolution of 'gift-presenting' behaviours by the male. The male of some empeid species presents the female with a prey item – some unfortunate creature captured by the male. While the female is distracted in the examination and consumption of the gift, the male mates with her and makes good his escape. The males of certain other species gift-wrap the prey item first, by encasing it in a silken sheath which they produce from silk-secreting glands. When presented to the female, she has first to unwrap the gift before she can consume it, and that takes time; the escape of the male is greatly helped by the additional breathing space that is bought with the gift-wrapping. Yet other species have evolved 'deceitful' male behaviours where the males cannot be bothered with hunting for a suitable prey item. What they do is gift-wrap a leaf or pebble, and so occupy the female with a phoney gift. Such deception reaches its height in a particularly cunning male, of a beast called *Hilaria sartor*. These males don't even bother with a phoney present. They just present the gift-wrapping, an empty silk sheath, and while the female is searching through it for something interesting to eat, the male mates and escapes – or tries to. This variation on a theme of giving gifts is an exquisite example of comparative evidence pointing to the evolution of behaviour.

ADAPTIVE BEHAVIOUR AS KNOWLEDGE

Running through all of this book is a single theme. It is that knowledge, as commonly understood, and adaptation are closely related. Indeed, if only because one has to begin understanding somewhere, and a simple and clear position is the best place to begin, I would like to get rid of the vagueness of phrases like 'closely related' and say that adaptation and knowledge are one and the same thing. Adaptations *are* knowledge. The implications of this position for our understanding of ordinary human knowledge, yours and mine – what it means for my saying that I know that Britain is a part of western Europe or that the boiling point of water varies with atmospheric pressure – can then be examined from this stark and unambiguous base. After all, our present scientific understanding of human knowledge is rather feeble. We can only improve on that, and one possible way to do that is to ground the study of knowledge more firmly in a science, evolutionary biology, that is better established than are any of the specifically human sciences like psychology.

The reader will recall that in Chapter 2 two particular features of adaptations were considered to be of special importance. One was their goal- or end-directed nature. Each and every adaptation is 'for' something. The three-dimensional structure of antibody molecules is 'for' its ability to bind with and combat a specific foreign substance; the wing markings of certain species of moth, which look uncannily like a large vertebrate eye, are 'for' frightening away potential predators; the webbing on the feet of a duck is 'for' increasing the thrust of leg movements in water and hence raising the efficiency of swimming. It is this being 'for' something, this purposefulness, that gives biology its teleological character and makes it so easy to talk about adaptations as having goals. It also leads directly to the second characteristic of adaptations. This is their relational quality. Every adaptation comprises organization of an organism relative to some feature of environmental order. The folding configuration of the antibody is the organizational end of the adaptation; it stands in relation to some specific antigen, which is the environmental side of the adaptive relationship. The wing markings of a moth stand in relation to the nervous system of a predator, specifically the way in which that nervous system is wired such that the 'eye' startles the predator and perhaps causes it to flee. And the webbed feet of the duck stand in relation to the density of water.

Fine, you say. But why take the further step of equating adaptations with knowledge? Why knowledge, of all things, when it has a well-accepted common-sense meaning and such a central and sensitive place in philosophy? How can the wing markings of a moth be knowledge? How can a lymphocyte be said to know something about a virus? Is this wilful misuse of a word, an excuse for an argument? Well, the answers to these questions have been partly given in the previous paragraphs. When you say that Europe is in the northern hemisphere, that knowledge comprises two components: a brain state, which is a part of organismic organization, and the world itself (or its depiction in maps), which is the feature of environmental order relative to which that brain state stands. All human knowledge has the same two-component relationship that adaptations have. Thus can the equation be tentatively made between human knowledge and adaptation. Furthermore, human knowledge is not divinely given – at least most scientists cannot accept that it is. Nor does it come out of nowhere; no biologist would accept that we have an ability that is without provenance in our evolutionary ancestors. It must have its origins in other phenomena in other living creatures. It must be connected to something else. The connection that I am arguing for is that *all* adaptations are instances of knowledge, and human knowledge is a special kind of adaptation.

Think of it this way. Adaptations would not be adaptations unless they had this end-directed property. This direction can only result if adaptations are 'in-formed' by features of the world; they are highly directed kinds of organization and not random, transient structures that may or may not work. Adaptations *do* work, and they work precisely because of this 'in-forming' relationship between organismic organization and some aspect of the order of the world. This 'in-forming' relationship is knowledge.

Four points complicate the matter and I briefly make, or remake, them here because no one should think that the evolutionary epistemological view of knowledge as being rooted in biology is a simple one. The first is in the form of a reminder that though living things and their component adaptations are highly unlikely entities and cannot be accounted for by some single chance occurrence – the eye did not leap fully formed into the world and neither did its possessors – none the less, at the heart of the evolutionary processes that lead to such improbable complexity are chance events, that is, the unguided or blind production of variation. That, of course, is the whole point of the notion of selection. It is an account of how improbable complexity is moulded out of chance variation. Even complex knowledge as commonly understood has chance variation at its heart.

The second point is that adaptations are seldom optimal, being constrained by the factors outlined in Chapter 2. They are more usually than not 'satisficing' rather than perfect. If adaptations are knowledge, then it follows that knowledge is only ever partial and incomplete. The relationship between internal structure and external order is never perfect. The 'language' of that relationship is also worth commenting on. In some cases the relationship may take the form of a kind of lock-and-key matching, where the knowledge contained in, say, the molecular configuration of an antibody is a partial isomorphic fit to the molecular configuration of the antigen. But in other cases the relationship is a more functional or utilitarian match. The webbed foot of the duck uses what tissue is available in the feet of ducks, but virtually any material impermeable to water and strung between the toes would do. Here the form that the knowledge takes is quite imprecise and should perhaps be expressed instead in terms of some index of utility.

Third, for simplicity I have mostly been assuming up to now that the adaptive relationship is always between an internal or organismic component on the one hand and some external or environmental component on the other. But this need not always be the case. Take the structure of a joint, like the human hip. Here the form of the pelvic and thigh bones are in adaptive relationship to one another, to the muscle insertions on the bones and to the requirements of a bipedal gait and gravity. In part it is a matter of where you draw the line between adaptive structures, and it might be argued that the entire joint including all bones and muscle insertions should be seen as a single adaptive unit. But such a unit is still just one end of the adaptive relationship. What is the other? The answer is not just gravity and the nature of the terrain over which walking and running occur and the fitness gains bestowed by an upright gait, but the actual behaviour of bipedalism itself, which is partly caused by neurological events. Again one can ask whether it isn't appropriate to include the neurological components as well as those of the hip joint to define an even larger adaptive unit, the bipedal gait, which will have many other components in addition to those just mentioned. Quite apart from any judgement that might be made about such a tendency to lump multiple attributes into single adaptive characters, such lumping still raises the same issue as keeping the component bits apart. This is that it may and does happen that adaptations involve not just internal–external relationships, but internal–internal ones as well. Such inward-looking adaptive relationships may be especially significant when one is considering the adaptive states of brain structures and their concomitant psychological adaptations.

It is worth repeating here too that when the adaptive relationship does involve internal–external components, that entity in the world outside which forms the external end of the relationship may be, and often is, another living creature. And a very special class of living creatures with which we are in adaptive relationship is members of our own species.

Fourth and finally, it must be remembered that neither the

inanimate world nor the living creatures within it are enduring and unchanging. Both change, and often the changes are wrought by the one on the other. And that includes the effects that living creatures have on other living creatures. So adaptations are not static things but dynamic in their interactions with the world. In the same way, knowledge must then be seen as dynamic and interactive, a view from which few psychologists would dissent.

That was a long digression from the main thread of the argument to which I want now to return. According to the earlier definition of Piaget's that we adopted, behaviour is a particular set of adaptations. That being the case, then behaviour is a particular kind of knowledge, even though that behaviour contains no necessary element of thought, reflection or memory. The behavioural differences between cliff-nesting and ground-nesting gulls can all be understood in terms of the relational quality of all forms of adaptation, and hence as differences between the gulls in terms of their knowledge of the world in which they live. The birds have brain states that are manifested in certain behavioural patterns – nest construction, for instance. The features of the world that are the external end of these relationships are either cliffs or level ground. The behaviour that results is accordingly different, and it is different because it represents knowledge states that are different. What the kittiwake knows about is cliff faces, and that is different from what, say, the black-headed gull knows about, which in this specific instance are the properties of level ground. That knowledge, in isolated behavioural examples, is of a very partial and fragmented form. The depth of cup of the nest is an exceedingly isolated and sparse form of knowledge of cliff faces by kittiwakes. But, of course, kittiwakes know more than just this about cliffs. Their total knowledge is much more impressive and covers the entire range of behavioural adaptations in the form of foraging, feeding young, defending themselves and their chicks, and so on. Taken together, these behaviours constitute a quite impressive knowledge of the cliffs, what might properly be called a working knowledge. This, to repeat a previously made point, is because knowledge is something that does work for its possessor. However, and I make no apologies for repeating this point too, just as adaptations can be thought of as satisfying, so knowledge has an equivalent in terms of deviations from perfection. Knowledge may never be absolute and certain, but it is always true enough to be workable.

The behaviour of the kittiwake is therefore a form of knowledge, and it is usually acceptable to phrase it in something like this way when one comments, 'Clever bird! It knows that it must forage in groups.' When making this kind of statement one does not imply conscious knowledge by the birds of the dangers of foraging singly and the advantages of doing so as a group. One merely looks at the birds and infers (or is told of) the goal-directed nature of their behaviour, and then it seems natural to think of them as 'knowing' about their goals. The behaviour of foraging, however, is no different in this respect from the shape of the antibody molecule. The antibody knows about antigen just as the kittiwake knows about the dangers of foraging singly. It only seems to be more acceptable to talk in terms of knowledge when referring to birds because one is talking about a large, warm-blooded creature that is seemingly doing sensible things – a bit like us, really. We are even more willing to ascribe knowledge to warm-blooded and furry creatures like dogs, and more again to chimpanzees. But each case, be it antibody or brain, amoeba or human being, is knowledge. What you and I have in our heads, well, that may be thought of as being very special. But though special, it isn't really different in kind. They are all forms of knowledge.

There is another 'level' to the story of behaviour as knowledge which we must now briefly consider. It concerns the role of genes and development. It has already been stressed that in biology something cannot come out of nowhere, and this applies to behaviour too. Behaviour is a particular set of phenotypic traits, and, like all phenotypic traits, its origins lie in a complex mixture of genetically coded information and its realization through development in environments lying within a certain range of conditions. We have already noted that behavioural geneticists have demonstrated that a very large number of behaviours have in part a genetic cause, and that such genetic coding and transmission are an absolute requirement if behaviours are to evolve as adaptations. Given the essential causal links between behaviour, genetics and development, the claim that behaviour is a form of knowledge should properly be extended to include the genetic and developmental factors that determine behaviour. Such considerations do not fundamentally alter the picture that I have been painting. Far from it: including them has two distinct advantages. One is that it gives us a more complete account of the nature of knowledge; the other is that it provides the essential – I would argue, the only – bridge by which we can cross from the evolutionary theory with which we began to knowledge as commonly understood. So we must now concern ourselves with the issue of genetics and development.

All sexually reproducing multicellular organisms like ourselves begin life at conception as a single-celled entity containing an array of genetically coded information (in the form of instructions for the building of proteins) and an energy store. Through the extraordinary, and as yet still poorly understood, cascade of processes that is individual development (often referred to as ontogeny in the technical literature), these genetic instructions become expressed as the highly differentiated multicellular creatures that we are. Development is not an automatic, pre-ordained unfolding process which, once initiated, proceeds to the completed state of the adult organism. Rather, each individual is, in a real sense, created anew, the unique outcome of an immensely complex series of interactions between the different parts of the genetic constitution of that individual; and also between its genes, its developing parts and its environment. Epigenesis is the word used to describe this complicated, integrated, dynamic and probabilistic process of development.

Because of the nature of epigenesis, one can never say that a phenotypic attribute is caused just by genes. The length of my nose was undoubtedly partly a consequence of my genetic make-up, but it was also in part a consequence of conditions of my development, the environments experienced. There is another, exceedingly important consequence of the complex and contingent nature of epigenesis. It is that my nose length might have been different had the conditions of my development been different. My genes specified some range of possible nose lengths. What I ended up with was in that range, but just where in that range was determined by other factors.

There are innumerable and well-documented examples of how variable the outcome of development can be. In the case of humans it ranges from the manner in which the binding of a baby's feet can reduce their size to the way in which lifelong self-esteem is determined by the experience of being loved as a child. Rather better controlled studies of the variable nature of the effects of development come from laboratory investigation into non-human development. Some of these investigations indicate just how blunt the effects of early experience can be, as in the cases of many species of insect that will fly well, poorly or not at all depending upon the ambient temperature during development. On the other hand, we know of very detailed changes that can be wrought by the conditions of development, none more famous (or infamous) than those of the locust. Whether locusts aggregate and swarm or live solitary lives depends upon the amount of exposure to other locusts early in their lives, and this in turn is determined by climatic conditions and the availability of food resources. Not only is their behaviour variable but so too is their general morphology – indeed, at one time locusts of the same species but in either swarming or solitary forms were thought to be members of separate species, such is the magnitude and intricacy of these developmentally induced differences.

Much is also known about how developmental differences induced by differences in experience can affect specific organs and organ systems. For example, the number of cells in those parts of the brain concerned with the analysis of visual input will vary with the amount of visual stimulation that the developing animal experiences. More specific changes in the visual system, such as the degree of sensitivity of the visual system to lines of a certain angle of orientation, can also be altered by early visual experience. Sexual determination is a particularly good example of how complex the interactions can be. All mammals during early development have a female morphology. The genes then determine whether or not a certain hormone is released, and it is this that determines whether there is continued development as a female, or whether male anatomy begins to appear. The male or female structures then cause the release of further hormones, which in turn act to affect which genes in other parts of the body are turned on or off, producing further anatomical and physiological changes. With such a complex pattern of development there is much scope for environmental effects to intervene and alter both the structure and the behavioural expression of sex differences, sometimes in ways whereby the effects are remote from the cause. For instance, in a classic series of studies carried out by American psychologists in the 1950s it was found that rearing monkeys with insufficient social interaction with their peers led to abnormal sexual behaviour as adults, even though sexual behaviours are absent in the young monkeys at the time that they were separated from their peers.

Epigenesis tells us something very important: that even though adaptive structures, including behaviours, must in part have a genetical cause, they are not necessarily invariant in form and may sometimes vary quite widely as a result of the environment in which development occurs. This developmental plasticity can itself be seen as an adaptive device, that is, as a knowledge-gaining device. If the coat thickness of a mouse varies with temperature during development, then not only is the adaptation of thickness of coat a form of knowledge of the temperature of the world of that mouse, but the relatively short-term process of development itself, together with the longer-term process by which the genes that control such flexible development are selected, can be seen as the integrated processes by which that knowledge is gained.

Let us be clear: the argument up to this point has been that knowledge has two components, one internal and the other external; that adaptations are to be seen in exactly the same way; and that all adaptations are forms of knowledge. But adaptations are usually features of developed, adult creatures. We must go further and think of them as the products of developmental processes. These developmental processes and the genes that initiate and participate in them should be seen as the integrated way in which knowledge is gained. It is possible to continue to think of this in terms of our two-component formulation. At conception the knowledge relationship is a potential one between the genes representing the internal end of the knowledge relationship, the external end being the range of possible developmental environments that have been the conditions exerting the selection pressures which led to those genes being selected over long periods of time. As development of the individual proceeds, actual environments and a reduced selection of genes in the total gene complement of that creature make up the external and internal components of the knowledge-gaining process. The end result is an adult organism with an array of adaptations, each an instance of knowledge with the usual matching relationship of organismic organization to environmental features.

So the argument is that knowledge – though not as yet knowledge as usually understood (which will be considered in the following chapters), but including behavioural knowledge of the kind that kittiwakes have about cliff faces – is a product of this complicated and dynamic developmental process. Before considering what is commonly meant by the term knowledge, however, we must examine the much maligned notion of instinct, and consider its limitations as a form of knowledge.

THE INSTINCT–RATIONALITY DICHOTOMY

I want to draw a distinction between behaviour that occurs without thought and which is not affected by the processes of learning and memory, and behaviour that is. For convenience and ease of understanding I am going to describe the former as instinct and the latter as the product of reason, intelligence, learning and memory – in short, rationality. In doing so, I am simply following the distinction that one first finds in the writings of Greek philosophers of half a millennium before the birth of Christ. There the use of the word 'instinct' was reserved for animals, which, it was claimed, could act only without reflection or thought – instinctively – whereas humans had rational powers to guide their actions.

This distinction, the instinct–rationality dichotomy, was, by and large, maintained for two and a half thousand years. The strength of the idea is attested to by its being held throughout this time despite humans' close association with domesticated animals such as horses and dogs and manifold examples of their ability to change their behaviour in response to training. Descartes was especially influential in propagating the dichotomy in recent centuries, emphasizing as he did the unthinking, brutish, machine-like behaviour of animals and contrasting this with the essence of the human soul, that is, rationality. Until the nineteenth century, though, there was no scientific theory which could accommodate, or even begin to accommodate, these two different 'kinds' of behaviour. The publication of Darwin's *Origin of Species* changed all that. Here, at last, was a vehicle that might eventually lead to a proper understanding of how behaviour that occurs without thought differs from behaviour that occurs as a result of thought.

As a naturalist, Darwin was conscious of the significance of behaviour to living things. What they did, how they acted, seemed to him to be every bit as important as their anatomy or physiology. And it was inevitable that his theory should lead him to think of behaviour at first entirely in terms of instincts, to which, as I have already said, he devoted a whole chapter of the 1859 book. His treatment of instinct was confined wholly to non-humans in that book, and it was only in his attempt to clarify the meaning of the word 'instinct' that he considered the case of human beings: 'If Mozart, instead of playing the pianoforte at three years old with wonderfully little practice, had played a tune with no practice at all, he might truly be said to have done so instinctively.' Darwin's meaning is quite clear. For him instincts are actions that are inborn, carried out without the necessity of practice or experience, and often in very young animals. They are also inherited and, in his view, are 'as important as corporeal structures for the welfare of each species', and the result of natural selection leading to 'the slow and gradual accumulation of numerous slight, yet profitable variations'. They were treated, in other words, in exactly the same way as all other adaptive characteristics.

Another way of saying all this is that for Darwin behaviour was to be understood in exactly the same way as the structure of a bone or the functioning of the kidneys. Selection has wrought gradual change on heritable variation to result in adaptive instincts. This was, and remains, a powerful way of dealing with the concept of instincts, and though he avoided the application of the idea of instincts to humans in 1859, in subsequent writings he was less reticent. What, though, of rationality? Darwin had very little to say about this part of the dichotomy in *The Origin of Species*, but more than made up for it in later work. A large part of *The Descent of Man*, published twelve years after *The Origin of Species*, was concerned with the 'mental powers of man and the lower animals', but the subject was handled in a very specific way. Darwin made little attempt to consider the function of rationality or what the relationship might be between rational powers and instinct. Instead his analysis was almost entirely a comparative one in which he was concerned to show that much of human rational powers has its beginnings in the rational abilities of 'lower' animals.

If no organic being excepting man had possessed any mental power, or if his powers had been of a wholly different nature from those of the lower animals, then we should never have been able to convince ourselves that our high mental faculties had been gradually developed. But it can be shewn that there is no fundamental difference of this kind. We must also admit that there is a much wider interval in mental power between one of the lowest fishes, as a lamprey or lancelet, and one of the higher apes, than between an ape and man; yet this interval is filled up by numberless gradations.

Darwin's main message was that, extraordinary as human rational powers may seem to be, they do not set human beings apart from their fellow creatures. Rationality is an attribute that we share with other species, and from whom we have inherited it. Continuity between species is all, whether it be a matter of instincts or rational abilities.

So the result of the rise of evolutionary theory was not the abandonment of the dichotomy of instinct and intelligence as separate causes of behaviour, but on the one hand a search for the beast-like, instinctive qualities of human beings, and on the other for the intelligent, rational capacities of animals. At first the latter aim was pursued through a weak, anecdotal method by which the wondrous mental capacities of beavers and baboons were imaginatively recounted by naturalists. But at the turn of the twentieth century, independently and half a world removed from one another, came the first controlled experimental studies of learning in animals. This was the work of Edward L. Thorndike in the United States, using cats escaping from puzzle boxes (referred to in Chapter 3), and I.P. Pavlov in Russia, studying the salivation responses of dogs, which led to the study of a form of learning which came to be known as conditioning. From these beginnings arose the important and exacting science of animal rationality.

We now know much about associative learning in certain animals, and we know also about their capacity for learning and remembering spatial positions and relationships. We know too about the ability of individuals of a limited number of species to reason by transitive inference and to be capable of other kinds of thinking. In sum, we now understand that certain kinds of animals can indeed think and learn, and this knowledge lies within the framework of a competent and growing science. The history of the search for human beings' beast-like qualities has, by contrast, been much less distinguished.

Darwin himself was rather cautious and uncharacteristically muddle-headed about the relationship between instincts and intelligence. He advanced the cases for both a direct and an inverse relationship. In the former, the more instincts and the more complex the instincts, the greater would be the animal's powers of intelligence; and in the latter, the greater the intelligence, the lesser the instincts. He seemed reluctant to take sides on the issue, but he certainly did claim the existence of a wide range of human instincts. These included very notably in his work the expression of emotions, but also some curious and vague ideas about 'social instincts' that included 'the love of praise and fear of blame'. One thing is clear: Darwin and most later evolutionists did not challenge the separateness of instinct and reason that is so starkly present in the original form of the dichotomy. Some behaviours are instinctive and occur without thought. Others are rationally determined. Instincts are one thing, and rationally determined behaviours are quite another. This is such an important point that I am going to give it a name: the doctrine of separate determination.

The turn of the century saw a burgeoning of instinct theory in psychology. Freud was an early instinct theorist who presented a serious and complex theory of human nature. Most others, however, did not, tending instead to 'explain' human actions, and even national attributes and historical events such as wars, by inventing the existence of a large number of instincts and giving them names like 'fear', 'self-abasement' and 'the need to dominate others'. Despite the conceptual emptiness of such an approach, the practice spread, and over the next twenty years or so hundreds of authors proposed thousands of instincts. My favourites, cited in a historical study of the period by R. Boakes, an English experimental psychologist, are 'the instinct of a girl to pat and arrange her hair' and 'the desire to liberate the Christian subjects of the Sultan'.

Such descriptive nonsense could not long pass for science, and the concept of instincts was swept away by the wave of environmentalism that flooded through the social sciences in the 1920s and 1930s. Psychology's most rigorous practitioners, the experimentalists, eschewed the notion entirely, and for several decades the concept of instinct, remarkably, had virtually no place within scientific psychology. However, behavioural scientists of a different ilk had begun to rise to prominence in the 1930s: the ethologists, led by Lorenz and Tinbergen, declined to reject the concept of instinct just because it had been misused by psychologists. The ethologists were biologists by training and steeped in the theory of evolution. The relatively stereotyped but adaptive behaviours that they observed, using strict methodology, in fish and birds in their natural environments cried out for understanding as instincts – that is, as adaptive behaviour patterns that are heritable and hence in part determined by genes.

The previous abuse of the notion of instinct was an uncomfortable burden that the ethologists had to bear, and during the 1950s the concept of instinct again came under serious criticism, mainly at that time from North American behavioural biologists and psychologists. Some of the criticisms were well founded, and others less so. An intense and, at times, rather bitter debate ensued out of which emerged a better idea as to how to think about instincts. It went something like this. An instinct is an adaptive behaviour or pattern of behaviours that is caused partly by genes and partly by the complex sequence of developmental events and experiences normally encountered by the individual members of a species. The result of such a complex genetic–environmental interaction is a brain that is structured to give rise to certain species-typical behaviours – the instincts. Because of the chance nature of such a complex interactive and dynamic process, the concept of instinctive behaviour does not imply that such behaviour is invariant in form, nor, obviously, that development is irrelevant to its shaping. What the notion does demand is that such behaviours, while not invariant, are none the less similar in form and may be widespread among the individuals of a species.

Now, in order to develop a particular line of reasoning, I need for the moment to keep such instinctive behaviours conceptually apart from behaviours that arise through processes of learning and intelligence. In the end we will refute this stringent and false assumption. It is false because the doctrine of separate determination is false. And the doctrine of separate determination is false – quite apart from the argument to be presented in the next chapter – not only because it is plausible that, once the abilities of learning and intelligence have evolved, such processes will enter into the shaping and determination of at least some instinctive behaviours in some species of animals, but also because there is good empirical support for such a view. However, unless we initially keep instinctive and learned behaviours apart we will not be able to understand why it is that rationality ever evolved in the first place. Knowledge as commonly understood is a product of the processes of rationality and if we are to understand what knowledge is, we must understand the origins and nature of rationality. In other words, we must be able to answer the question *why all behaviour isn't instinctive*. That is, *why did rationality ever evolve at all?* We can only answer this if we impose the instinct–rationality dichotomy as an analytical device.

The question of why all behaviour is not instinctive, of why rational powers ever evolved in the first place, concerns the ultimate causes of rationality. Remarkably, it is an issue that has attracted little attention until recently. Darwin never put the question in this form, and for a hundred years almost none of his followers did either. The beginnings of this approach to the evolution of intelligence can be traced to the 1960s, when the application of cost–benefit analysis to an understanding of adaptations began to spread in behavioural biology.

Cost–benefit analysis, in a nutshell, is a way of thinking borrowed from economics which considers every facet of living organization, from obvious bodily adaptations to the amount of DNA in a cell, in terms of the costs and benefits associated with it. While related to the traditional measure of fitness in terms of reproductive competence, the currencies of costs and benefits are often very abstract, but in general the nature of the approach is easy to understand. For example, being fleet of foot carries with it the benefits of avoiding becoming a meal for some other creature or enhancing one's own chances of capturing prey. The costs are those of developing and maintaining the necessary skeletal structures and musculature that will allow for fleetness of foot, together with the metabolism that will support massive energy expenditure over short periods of time. If the benefits outweigh the costs, then an attribute like fleetness of foot will evolve and be incorporated into the organization of a living system.

Using this approach we can ask what are the relative costs and benefits of performing a task through behaviours that are instinctive as opposed to the costs and benefits of performing those same behaviours when they are controlled by thought and memory. In order to begin to answer this question, we must briefly list some of the characteristics of rationality to which we will return in the next chapter.

One of the most important is that rationality, intelligence, thought and the ability to learn and remember are never open-ended *tabula rasa* abilities. Rather, they are constrained, limited and directed. Learners can only learn and think about certain things, not about anything. This means that for an instinctive behaviour the instructions that have to be carried by genes, and the instructions that later have to be carried in the brain and the computations that have to be carried out by the brain to produce the behaviour – say, the recognition and care of young – will be fewer than those required for that same behaviour to be acquired and controlled by learning and thought. This is because the instructions for constructing a nervous system that will perform the necessary sensory and motor tasks will be the same in each case – recognizing young and feeding them require the same recognition processes and the same motor actions, whatever the cause of the behaviour. But in the case of learning, there will have to be, in addition, dedicated neural networks that will carry out the necessary computations that result in the appropriate stimuli and responses being learned, together with some kind of storage-and-retrieval mechanism. And these additional brain requirements will have to be specified genetically and emerge from the developmental process.

The instructions, in other words, are divided between genes, development and functional brains, but in the case of instincts the costs are less both in the currency of genetic information and in the amount of brain given over to them. The brain is metabolically the most expensive organ in the body; so, reasoning in this way, behaviours that are instinctive are 'cheaper' than those that are caused by learning and thought.

There is another consideration to take into account. Instincts, like all other phenotypic adaptations, have been formed by a long process of trial and error; instinctive behaviours, such as following a mother or avoiding dangers, are usually 'correct' because the adaptive fit between them and the environmental conditions with which they are in adaptive relationship have been honed and sharpened over hundreds or thousands of generations or more. What we see now are the 'good' instincts selected by this slow process acting over a long history. Learning, on the other hand, is more error prone because the fast-acting learning process can always be derailed by momentary and chance events which force mistakes. This may not happen too often, because learning, as we will soon see, is primed by genes. Learning none the less can and does make errors more often than the more plodding but certain processes that form instincts. So, in this sense, instincts are also more reliable than learning in causing or determining adaptive behaviour and hence should have greater benefits.

And yet, despite the apparent superiority of instincts over rationality, the latter did indeed evolve. This *must* mean that rationality scores points over instinct in some creatures such that the balance of costs and benefits has shifted from favouring instinct, or at least some instincts, to intelligence. We can now turn to the limitations of instinct, and how rationality makes good these limitations.

SUGGESTED READING

Eibl-Eibesfeldt, I. (1975) *Ethology: The Biology of Behaviour*, 2nd edn, New York: Holt, Rinehart & Winston. This book certainly does not represent a contemporary view of ethology, but it has an interesting chapter extending the notion of instincts to humans, using the framework of classical ethology.

Gould, J.L. (1982) *Ethology: The Mechanisms and Evolution of Behaviour*, New York: Norton. A useful account of the modern study of animal behaviour, by an author unafraid to call an instinct an instinct.

Lorenz, K.Z. (1958) 'The evolution of behaviour', *Scientific American*, vol. 199, no. 6, pp. 67–78. A simple account by the founder of classical ethology.

The Evolution of Intelligence

===

Behaviour without thought – instinct – serves a very large number of animals well. Estimates as to how many species existing today are able to supplement their instincts with some additional device for shaping adaptive behaviours will differ widely depending upon where one draws the boundary between instincts whose form may vary because of developmental flexibility and behaviours that are shaped by learning and/or thought. Nor does it matter much to the arguments that follow in this chapter exactly where that line is drawn and hence what number of animals fall on the one side of it or the other. My own, perhaps extreme, view is that only about 5 per cent of extant species do supplement their instincts with these additional devices for shaping adaptive behaviours. If one expands that estimate to consider *all* species that have ever existed on Earth, it would be a fraction of one per cent.

Much the greatest number of species are, and have been, invertebrates (animals without backbones), and the majority of these are insects. No one knows how many of these have only instinctive behaviours to serve their needs, but my guess is that the answer is most of them. That, however, merely underscores how successful an animal can be when its behavioural adaptations are solely instinctive. None the less, for a significant minority of animals – including mammals and some other vertebrates, and also insects such as bees, and some molluscs such as squids and octopods – instincts alone are not enough. These are species whose 'life-styles', technically referred to as life-history strategies, are such as to bring out the limitations of instincts.

In truth, this is a limitation confined not just to instincts, but to all forms of adaptation, to all forms of knowledge. This limitation is one of the fundamental problems that every living creature faces when confronting a dynamic and changing world, equipped only with the adaptations furnished by genes and development; in this book, however, the problem will be discussed largely in terms of instinct. In order to understand what this limitation is, we must go back and restate in a very truncated form where the instructions for constructing adaptations come from.

THE UNCERTAIN FUTURES PROBLEM

The gene pool is the sum total of all the genetic material within a breeding population, and as such includes all the genetic instructions for the building of adaptations in individual members of that population. While an abstract notion, the gene pool is also a very useful one and so worth dwelling on briefly, because it will help us to get clear just how an animal comes to be furnished with an array of adaptations. One must remember that these adaptations are constructed on the basis of genetic instructions that, through a complex process of development, become expressed as organic structures that relate to some feature of environmental order.

Now, the reason for the concept of a gene pool being abstract is because, contrary to the idea conjured up by the word 'pool', which usually means some single thing in a fixed place, the total genetic material within a breeding population is actually something that is divided up and scattered about among the individuals in a population. Being a part of the constitution of all the reproductively competent individuals in that population, it certainly is not some single thing in a fixed position – indeed, it may, and does, move about. It is an abstraction for another reason too. The very word 'pool' conjures up a vision of homogeneity, of a lack of structure. But, in sexually reproducing creatures like ourselves, offspring are formed from the genetic material of two parents. A lack of mating structure within a gene pool would mean that mating was random. Mating, however, is usually not random in breeding populations. Witness, for example, just how common are dominance hierarchies in primate species where dominance is a significant determinant of which animal mates with which. Or consider the harem structure of many species of deer, where single males have access to many females, jealously guarding them from the attentions of other males. And who mates with whom in a sexually reproducing species is very important. This is because it is not the case that all genes in the genotype of an individual are always expressed; and those that are expressed are not invariably expressed in the same way. The conjunction of genetic material from two parents, and the interactions between genes that result, are crucial in determining which genes are expressed and how.

So the idea of a gene pool certainly is an abstraction and involves some simplifying fictions. None the less it is convenient as an image, because one can imagine an individual at conception 'dipping' into the gene pool and coming up with a 'handful' of genetic instructions. Such a 'handful' contains only a very small sample of the genes and combinations of genes that are available in the entire gene pool, but is none the less the total genetic complement of that individual. What is more, that is all that the individual is ever going to get in the way of genetic instructions from the gene pool for the rest of its life. It must somehow go from conception through to adulthood with just those genetic instructions and no others. When it becomes reproductively competent it will return those instructions to the gene pool if it is reproductively successful, but not if it is unsuccessful — in which case, that particular sample of genes will be eliminated from the gene pool. In summary, each individual begins its life with a small sample of genes drawn from the much larger total genetic constitution of the population of which it is a part; it then expresses those instructions in organic form as a result of development; and then it may return those genes to the gene pool at some later time when it has (or has not) become reproductively capable.

At some moment in time, then, let us call it time$_1$, a creature is furnished with a set of instructions for building an adaptive, competent phenotype. Those instructions must, as discussed in the previous chapter, be expressed by a process of development that takes some period of time. At some later time, time$_2$, our organism may (or may not) be in a physiological state allowing it to enter into reproduction and hence to return its genes to the gene pool. This whole process takes a finite period of time, which is the difference between $time_2$ and $time_1$. In animals like us humans, that is a period in excess of ten years. This period of time will not matter if the state of the world does not change for the animal during that time. In other words, the instructions received at $time_1$ will still be appropriate at $time_2$. But if the world has changed in that time, and changed in ways that are significant for the animal, then the instructions received at $time_1$ may be out of date and no longer appropriate at $time_2$.

This period of time between receiving a set of genetic instructions and their implementation, through development to the point where those selfsame instructions might be returned, via reproduction, to the gene pool, has been given a variety of names. My favourite is that of Konrad Lorenz, who called it generational deadtime. It is a lag-time that is an invariant feature of any system whose construction takes time and which is based on a set of instructions that cannot be continually updated. In the case of sexually reproducing organisms, each organism is cast off and cut off from the gene pool with a fixed set of instructions that cannot be altered. It cannot 'dip' back into the gene pool to augment those instructions if it finds they are not good enough. In this sense, genetically, it is on its own.

In an unchanging world this is not a problem. Consider the analogy of an architect's plans for a house. They are a set of instructions that apply at the time the architect hands them over to the builder. Lacking the ability to alter the plans, the builder just builds to their specification. If the weather does not change, and if the ground level remains constant, then the plans will result in an adequate house. But let us assume that there is significant change – say, ground subsidence – after the plans have been handed over. And to make the analogy more complete, the builder *cannot* simply alter the plans by referring back to the architect. The result is a defective house. So, in a changing world, the time-lag may be a very serious problem. What, in effect, nature does is consign organisms to a future about which there can be no certainty, with

a set of instructions for the building of adaptations which may no
longer be right for the job. This danger of always being 'out of
date' in one's genetical knowledge of the world must somehow be
countered.

C.H. Waddington, the British biologist, put this wonderfully
well in a piece written in the 1960s: 'The systematic exploration of
the evolutionary strategies in facing an unknown, but usually not
wholly unforecastable, future . . . take(s) us into a realm of thought
which is the most challenging . . . of the basic problems of
biology. The main issue in evolution is how populations deal with
unknown futures.' How, he asked, does life maintain itself on the
basis of potentially inadequate information?

THE LOGIC OF THE G-T-R HEURISTIC

What Waddington was really asking about was change and how
life copes with change. If we are to answer his question we must
consider both the nature of change and, very importantly, the
'logic' of provisioning organisms with instructions only via genes
and the developmental processes through the medium of the g-t-r
heuristic described in Chapter 3. (Because the phrase 'genetic-
developmental g-t-r heuristic' is awkward and ungainly, I am
going to refer to it as the primary heuristic. The reasons for the
'primary' will soon be apparent.) Now, remember that the g
phase is the generation of variation by way of the variety of
mechanisms described in Chapter 2, such as mutation and the
independent segregation of chromosomes during the formation of
sex cells, as well as that stemming from the variable outcome of
epigenesis described in Chapter 4. The t phase is selection, which
determines which variants will be fed back into the gene pool and
so propagated in time. The r phase is the regeneration of further
variant forms, some of which are the conserved forms of previously
generated variants and others of which are novel, resulting from
those same and ever-present variation-generating mechanisms,
including mutation, generation of novel combinations of genes
and chromosomes and novel developmental pathways.

The primary heuristic has two particular features that are worth noting. One is that it takes the logical form of induction, generalizing into the future what worked in the past. That is, the successful variants are fed back into the gene pool where they will be available for sampling by future organisms. This is the conservative, pragmatic part of the heuristic. The other is the generation of novel variants by chance processes. This is the radical, inventive component of the heuristic. It is nature's way of injecting new variants into the system in order, possibly, to make up for the deficiencies that may occur if what worked in the past no longer does so because the world has changed. When John Odling-Smee and I first wrote about this in the 1970s, we noted: 'In effect the g-t-r heuristic "gambles" that the future will be the same as the past. At the same time it hedges its bets with aleatoric (chance) jumps, just in case it is not.'

What, then, of this matter of change in the world? How can one best think about it? This is not an easy question, because change is a universal condition of the world. If the world were unchanging, then evolution would have proceeded to some optimal point and then ceased. This has not happened. Nothing stands still, and the very occurrence of evolution is both a force for change itself and proof positive of its existence.

Given the pervasive nature of change, perhaps the best way of beginning to answer the question is first to consider how change can be categorized in some simple manner. One way of doing this is to divide change into what can be predicted and what cannot. Prediction here refers not to what scientists armed with the tools of their trade, which allow them to escape from their own individual perceptions, now are capable of predicting. The predictor that we are considering here is a localized observer, such as a beetle buried under the bark of a tree, a bird in the forest or a lizard on the ground.

Sudden, perhaps catastrophic, change cannot be predicted by such an observer. Such change cannot be tied to any preceding events by the localized observer who is affected by it, and it is often followed by a long period of stability. Dramatic examples are meteorite strikes, earthquakes and a sudden invasion of a new pathogen. Because they are unpredictable, such changes cannot be guarded against by living systems except by way of the chance or radical component of the primary heuristic. Organisms either just happen to have the right features and survive, or they do not. It is precisely because such changes are sudden, effectively without precedent, that the inductive logic of the conservative component of the primary heuristic cannot supply effective adaptations to cope with them. Only if there is an element of repetition, of predictability, of projection into the future of what occurred in the past, can the primary heuristic work inductively as a supplier of adaptations, that is, as a supplier of knowledge. Unpredictable change is a form of the uncertain futures problem that cannot be solved in this way.

Yet sudden and unpredictable change need not be complete and need not be catastrophic. For a beetle in a tree, the blowing down of that tree may or may not be as catastrophic as an earthquake. And if it survived, sudden change may not result in a future that is entirely different from the past. Indeed, it is unlikely to do so. The invasion of a habitat by some ground-dwelling predator may prove a fatal problem for certain creatures, like mice, that are mostly restricted to a ground range. But for a squirrel able to survive much of the time in trees, though the change is serious and it must be on its guard when on the ground, the trees are still there and so too are innumerable other features of its familiar environment. None the less, the extent of sudden change is no more predictable than the occurrence of the change itself, and how well an organism can survive in an unpredictably altered environment is a matter of chance.

What then of predictable change? The easiest way of saying what this phrase means is to refer to the most obvious example of it, namely the cyclic alterations in temperature and amount of light, as well as a host of related changes, that make up the cycle of the seasons. The cycle itself, of course, is unchanging, or relatively unchanging. But it is probably safe to assume that only we unusually perceptive *Homo sapiens* can know this through individual experience. It is within any one cycle that change occurs, and our localized observer, be it a beetle, bird or lizard, experiences the cycle as change. If, as we shall see in a moment with a specific example, the life span of our localized observer is one year or less, then that organism cannot possibly acquire knowledge about the changes that make up the cycles through its own experience. Moreover, it is extremely doubtful that those animals that do experience several such cycles learn the sequences of changes either – the duration of time over which such events have to be integrated almost certainly takes it beyond the conceptual abilities of even the cleverest of apes. However, there is no doubting that animals and plants are adapted to the seasons. How has this happened? What is the predictor that allows for the evolution of such knowledge which is beyond the individual cognitive capacities of any single creature apart from ourselves?

The answer is that it is the gene pool, the accumulated genetic knowledge of whole populations acquired by way of the primary heuristic that is the predictor. It can fulfil this role because the changes that are contained within the cycle are repeated, and repeated well beyond the generational deadtime of any species of organism. This endless repetition of events in the past has resulted in the gene pool gaining by the inductive logic of the primary heuristic the genetic instructions that are transmitted across generations and hence projected into the future as adapted organisms. It isn't the individual that knows inside its head about the changes that characterize the cycle. It is the gene pool that knows and the knowledge is expressed as adaptations that fit the animal to the seasons.

Consider an example. In northern Europe, many species of ground beetle are conceived in the summer when these animals lay their eggs. Because of the impending weather changes associated with the autumn, the eggs are laid in a protected underground environment; thus, when the larvae emerge from the egg during the autumn, they are able to cope with these conditions which are different from those of the summer. The larvae, after a period of rapid growth, then pupate, a condition that affords them further protection from the rigours of winter, when again the conditions are different from those of the earlier months in the beetle's life cycle. The adults emerge in the early part of the summer, and the eggs of the next generation are fertilized. This complicated life cycle is tuned to the cycle of the seasons, and this tuning is an adaptation. Specific environmental inputs lead to hormonal states that trigger the succession of larval and pupa phases. The structures that detect these environmental events and the links to hormonal changes are determined partly by genetic instructions. The knowledge for these structures resides in the gene pools of the species, having been gained by the accumulation of appropriate genetic instructions over a long history of selection – that is, by the operation of the primary heuristic acting on genes and developmental pathways.

So predictable change means change which has occurred with sufficient frequency in the past that the processes of the primary heuristic have been able to accumulate the genetic instructions that furnish the adaptations that cope with such change in the future. Now, we know through our scientific knowledge that such cycles do indeed contain precise and repeated features, but such astronomical details aside, the cycles, while predictable, are actually experienced by our localized observers as rather 'wobbly'. Such 'wobbliness' can be referred to as predictable unpredictability – it is what Waddington meant with his phrase 'not wholly unforecastable'. Predictable unpredictability is the core concept for an understanding of why intelligence, rationality in any of its forms, evolved; it is the condition of the world that instincts as adaptations cannot deal with. Again it is most easily understood through an example.

Exactly what the temperature will be during each day in any three-month winter period is impossible to predict. The most skilled of meteorologists using the largest available computers cannot predict it, much less a squirrel that has to survive the winter out in the open. But it certainly is a very good bet that the average of those winter temperatures will be less than those of the preceding summer and autumn months. It may, of course, by a freak of weather conditions, turn out not to be so, but the chances are against this, and most of us would bet large sums of money on it. So, while I cannot say exactly what the temperature on this coming Christmas Day in London will be, I can predict with some confidence that it will be colder than it was on 25 June when people were out of doors watching the tennis at Wimbledon, or doing other summer-time things. This is an instance of predictable unpredictability. Winters are colder than summers, but we cannot say much about the fine-grained pattern of the temperature on any one day. Now, what I know through direct experience as well as through the reading of elementary meteorology texts, so too does the squirrel, but through its genes. It hoards food and grows longer fur in accordance with genetic instructions that 'know' that winters are colder than summers. The knowledge, and the prediction based on it in the form of fur lengthening and food hoarding, is coarse. It is none the less roughly correct. Like the man from the Meteorological Office, the squirrel cannot know precisely what the temperature will be on any one day. But what the squirrel knows is good enough, because it works. It is good working knowledge.

The wobble on the temperature cycle that characterizes the seasons is very fine-grained. Consider another example, where the wobble is rather more like a series of steps and plateaux, though ones attached to some predictable feature of the world. For instance, a child cannot be equipped by its genes with knowledge as to who in the future will be kindly and protecting and who will be threatening and hostile. What the child can be 'told' by its genes is that facial characteristics are a sensitive and significant feature by which people can be distinguished from one another. This is the predictable part of the predictable unpredictability formulation, and that is why we are equipped with a phenomenal ability to distinguish between, and remember, a very large number of faces. But, of course, the unpredictability concerns exactly what the facial characteristics will be of those who are kind and those who are dangerous. So here too we can reasonably describe the child's situation as being one of predictable unpredictability, but in this case there is a much coarser grain to the unpredictability.

From our chosen examples, predictable unpredictability can usefully be thought of as coming in two forms. In the one, as in the case of fluctuations in temperature, the temperature itself and the time for which some specific temperature is maintained are both unpredictable. These changes are effectively random – though, of course, conforming to the predictable cycles of the seasons. And in general, the time for which some single temperature is maintained is short – usually only the very tiniest fraction of the total life span of most multicellular organisms. Such exceedingly brief and changeable temperature levels might be described as an unstable feature of the world superimposed upon the long-term stability of seasonal flux. The states of mind underlying the behaviour of people, by contrast, are usually more stable. They are not constant, but friends tend to be friends for relatively long periods, and so too with foes. It may not be long-term in the way that the cycles of the seasons are long-term and permanent features of the lives of any animal, but friendship and enmity are more stable than the daily or hourly levels of temperature. So we can describe the attitudes and commitment of people to one another as a form of short-term stability. Such short-term stabilities are in turn superimposed upon a long-term stability, which is a person's face. We can change our behaviours, but we cannot alter our faces. Remembering these distinctions between instability, short-term stability and long-term stability will be helpful to understanding the next section of this chapter.

SOLVING THE UNCERTAIN FUTURES PROBLEM

The uncertain futures problem concerns an organism going through life, equipped only with instructions given at conception (and hence perhaps only correct at that time) on how to survive, and having to interact with a world that may be different from that in which its life began. There are, of course, two possible outcomes – the problem is solved or it is not. The latter is much the most common. If organisms cannot cope with the changes in the world, then they will die because of those changes. If all members of a species suffer the same fate, then that species will become extinct. Extinction is always caused by an inability to cope with the uncertain futures problem, and given the estimates that well in excess of 98 per cent of all species that have ever existed are now extinct, it is clear that, in the end, all life succumbs to its uncertain future. Extinction seems to be the universal fate. So really the second outcome is less a solution, more a struggle against the uncertain futures problem, a matter of delaying the time at which organisms are overwhelmed by their uncertain future. None the less, this struggle to delay the inevitable succumbing can be relatively successful. There are, after all, species that have existed for millions of years.

There are several ways by which the problem of change can be confronted, and none is exclusive of the others. One set of ways is for an organism to reduce the amount of significant change it has to deal with – in other words, reducing as far as possible the degree to which the genetic instructions of each individual are likely to be out of date by the time reproductive competence comes around. And the most common way of doing this is to reduce the period of time between conception and reproductive competence, and perhaps buttressing this strategy by producing large numbers of offspring, some of which may, by chance, survive in a changing world by eking out an existence in isolated pockets of little change. This means that the ratio of life-span length to numbers of offspring is low – that is, there is a high reproductive output in a relatively short time. This is a characteristic 'life-style' of animals known to ecologists as r-strategists (after the designation of selection forces first put forward by the American biologists R.H. MacArthur and E.O. Wilson in the 1960s). These r-strategists usually live less than one year, often considerably less than that; they develop rapidly; they are usually of a small body size; and they normally reproduce over just a single period. This is the life-history strategy common to most invertebrate animals and clearly one that is a relatively successful means of staving off the eventual victory of the uncertain futures problem.

Another way of reducing the amount of change in one's world, especially when much of that change is generated in one form or another by other creatures, is to live in a relatively isolated and unpopulated place, like a deep-sea trench or the top of a high mountain. The obvious difficulty here is that such niches are relatively empty precisely because they are so hard to live in. Very specialized adaptations have to evolve in order to deal with the great pressure and temperature difficulties of life at the dark bottom of an ocean. A variation on this theme is for parents to reduce change by protecting their offspring from the elements and predators, and by provisioning them with food – all the comforts of home life, whether it be a chamber excavated underground by a wasp or the nest of a bird. Such devices for the provision of home-comfort are themselves complex adaptations driven by the selection pressures of the uncertain futures problem.

The second very general way of dealing with change takes the form of 'if you can't beat it, join it': that is, change the phenotypes so that they can change with and match the changing features of the world. This, in turn, comprises two general strategies. The one results in changes *between* phenotypes, and relies on the chance or radical component of the primary heuristic. Again, large numbers of offspring are produced, but in this case each is different from the others. This broad range of phenotypic forms may result in some individuals, by chance, possessing adaptive features that lead to survival in the face of sudden and unpredictable change. This may be combined with one or other of the strategies to reduce change, and it is worth stressing again that none of these strategies need be exclusive of others. Indeed, the general rule has almost certainly been to combine ways of surviving change. For example, r-strategists – short-lived and hence less likely in the case of any one individual to have to deal with sudden alterations of the world – often do combine a reduced life span with a quite prominent radical component of the primary heuristic, so that even within a reduced uncertain futures problem a range of phenotypic forms is produced, some of which may prove capable of withstanding change. And they may further combine both of these strategies with yet another means of reducing change by providing some form of parental care, such as provisioning emerging young with a food supply in a protected environment. Many species of insects do exactly this, by laying eggs in enclosures in the bark of trees, for example. Thus they use a three-pronged defence against the change that might impinge upon their young. Insects, of course, are an astonishingly successful group of organisms, and this is almost certainly because of the good results achieved by such multiple strategies in combating the uncertain futures problem.

Another form of the 'if you can't beat it, join it' strategy is to match change with change, not by producing different phenotypes of fixed form, some of which may match the altered conditions of the world and hence will survive, but by giving rise to change *within* phenotypes: that is, by producing phenotypes that change in response to changes in the world. I will call this the tracking option.

It is self-evidently the case that the tracking option can only work when there is some element of predictable change, perhaps one in which change occurs within certain predictable limits. This is for two reasons – first, because tracking as a process requires knowledge that tracking is what must be done, and the knowledge for this can only come from the primary heuristic and be stored in a population gene pool. In other words, tracking involves mechanisms that must be put in place by the primary heuristic, and this can only occur when consistent, predictable selection pressures are present. Second, the primary heuristic must prime these mechanisms to track only certain things and not others. This, in turn, is because an all-purpose tracking device would, biologically speaking, be impossible to construct. It would have to have a sensitivity to all possible forms of energy changes in the world and be able to counter such changes with any response imaginable. No living creature has that kind of generality.

The tracking option itself can occur in two forms. Each is an appropriate way of dealing with that 'wobble' on predictable change, that is, predictable unpredictability. In the one case the tracking devices have evolved to deal with a wobble that is so fine-grained as to constitute unstable change. In the other the wobble leads to short-term stabilities. As an example of the former we can consider temperature, which fluctuates around some predictable seasonal cycle. Most animals are able to track such minor fluctuations and to respond in a variety of ways in order to maintain their body temperatures within some tolerable range. Temperature-sensitive sensors have evolved whose output determines certain physiological responses such as shivering, sweating, panting and presenting large surfaces well supplied with blood vessels (like the ears of elephants or the spines of certain dinosaurs) to the cooling air; or a number of behavioural responses have evolved, like moving into or out of direct sunlight or seeking shelter from wind chill; or there is some combination of both. Another example of such tracking devices is the light-attracted behaviours of such insects as moths that lead them into the 'erroneous' behaviour of circling and contacting artificial lights. These behaviours evolved in order for the insect to maintain an appropriate orientation to light when the animal is capable of movement that under some circumstances may take it away from that light. Indeed, most behaviours have these 'steering' components whereby behavioural output is guided relative to certain anchoring reference points such as gravity or a gradient of chemical concentration. Such reference points, and orientation by them, are essential when behaviour allows relatively unfettered movement in space, such behaviour itself generating a large amount of transient change for the animal.

Whether these ways of tracking high rates of change are physiological or behavioural, the machinery by which such tracking occurs does not itself change state during the act of tracking. Furthermore, it bears repeating that such machinery is an adaptation, or set of adaptations, that is there because it has evolved through the operation of the primary heuristic. The knowledge for such devices lies both in the gene pool and in the fit between the adaptive structures themselves and the features of the outside world that change so rapidly and to which they relate.

The second kind of predictable unpredictability, remember, comprises a wobble that is rather longer-lasting – what I previously called a short-term stability. These too are rapid changes, where features of the world may change several or many times in an animal's lifetime. For this reason they cannot be 'seen' by the primary heuristic. None the less, they may be of great significance for the animal, and the features may remain unchanged for a considerable length of time. The location of a good feeding site; the position of a bird's own nest and young; who one can rely on for support; who is dangerous: all these are of great importance and yet may, and often do, alter many times in the life of an animal.

How can such changes be tracked? The only way to do it is to evolve tracking devices whose own states can be altered and held in that altered condition for the same period of time as the features of the world which they are tracking hold fixed values. In other words, animals must evolve additional knowledge-gaining devices whose internal states match those features of the world that we are calling short-term stabilities. Such tracking devices would be set in place by the usual evolutionary processes of the primary heuristic and hence would operate within certain limits. But the exact values within those limits that these devices will settle to, and for how long, are not within the power of genes to decide. So devices such as these have a degree of autonomy in their functioning that makes them partially independent of both genes and development. These additional knowledge-gaining devices comprise a second – or, as we shall see in a moment, more correctly a secondary – heuristic.

In fact, there are two such classes of semi-autonomous knowledge-gaining devices or secondary heuristics that can track change in this way. The immune system is one; the intelligence mechanisms of the brain are the other. The immune system is the means by which the problem of the uncertain chemical future is solved. The brains of animals that have learning and memory mechanisms, or some further capacity for rationality, are the way in which the problem of the uncertain physical future is solved.

The situation with regard to the brain can be set out as follows. Think of the world as comprising sets of features, some unchanging and others changing at different rates and with different degrees of regularity. Irregular and infrequent change is unpredictable, and only sheer luck – somehow providing appropriate adaptations either through managing to avoid such change by being in the right place at the right time, or in the form of the radical component of the primary heuristic – will govern survival. What of more regular and frequent change? Well, if the frequency of change is less than that set by generational deadtime for extracting genetic instructions from the gene pool and then returning them to it, then the conservative component of the primary heuristic will be able to 'see' these changes and will furnish adaptations to match them. But if the frequency of change is faster than the frequency set by generational deadtime, then though the primary heuristic will be able to see the long-term stabilities upon which these changes are superimposed and will be able to detect the margin within which these more rapid changes occur, in order to track the precise values of these changes the primary heuristic will have to evolve devices that operate at a much higher frequency – at a frequency high enough to be able to track these values. If the high-frequency changes are unstable, the tracking device itself does not need to change state. Indeed, it would be maladaptive to do so. All it need do is command an immediate compensatory response. However, if these changes also have temporal characteristics that make them short-term stabilities, then these tracking devices must comprise a secondary heuristic that is able to change and maintain new states that match those features of the world that are being tracked. Such brain mechanisms – this secondary heuristic – that match the physical changes of the world are what we know variously and collectively as rationality or intelligence.

In addition to claiming that all behaviour should by definition be seen as adaptive, Piaget also used to like to say that intelligence is an adaptation. I have now presented a quite specific and simple picture that explains the general nature of that adaptation. Intelligence is an adaptation that allows animals, including ourselves, to track and accommodate to change that occurs at a certain frequency. Slower rates of change are adapted to by the genetical and developmental machinery. Higher rates of change are adapted to by tracking mechanisms whose own states are not altered by the transient events to which they respond. Intelligence, rationality, is an adaptation that has evolved to deal with changes that occur at rates somewhere between these two.

It is very important to remember that what constitute instability, short-term and long-term stability are all relative to generational deadtime. There is no simple and absolute identity for all animals of some range of frequencies of change that comprise long-term stability, nor others that make up short-term stability and a third making up instability. The same features of the world changing at some fixed absolute rate will constitute different forms of change for individuals of different species. The change from autumn to winter will be a gradual linear change for a bug that lives only a matter of weeks, part of an unrepeating sequence of change for a bug that lives for one year and part of an oft-repeated cycle for a bug that lives for many years. The nature of the adaptation that will evolve to match that change in temperature will vary accordingly. The same consideration applies to the response to change when that change involves individuals. A short-lived bug may not have time to enter into co-operative relationships with others. All that it must do is recognize a member of the same species and mate, all of which can be achieved by the kinds of tracking devices that themselves do not alter state as they track. Built-in recognition and steering mechanisms will do the trick. But it may be to the advantage of a longer-lived bug to recognize others and to form longer-term relationships with them — but not necessarily for the rest of their lives. In this case the tracking devices must be what we have now defined as intelligent.

In general, the kinds of animals that would evolve intelligent adaptations are relatively long-lived and produce relatively few offspring in their lifetime; that is, the ratio between the length of life and number of offspring is high. Such animals usually develop quite slowly and have a relatively large body size. They are what ecologists refer to as k-strategists. Some vertebrates are k-strategists. All primates are k-strategists. Humans, of course, are particularly long-lived and slow-developing k-strategists: it should not be surprising that some special kind of intelligence, some unique form of rationality, should have evolved in the case of our species.

Two general points must be made clear. First, none of the arguments presented here for the evolution of strategies for coping with the uncertain futures problem is dependent upon group selection. Individual or gene selection is all that is needed for the primary heuristic, working with consistent selection pressures over long periods of time, to evolve any of the strategies discussed.

Blind variation of genes and gene combinations that code for the complex adaptations that underlie these strategies are as capable of evolving immune systems and intelligent brains as they are of evolving the eyes of vertebrates or the complex social behaviours of bees and ants. Group selection is not envisaged as having any part to play in the evolution of strategies for dealing with the uncertain futures problem.

Second, though the problems are those of dealing with an uncertain *future*, no suggestion is made, and no possibility exists, that any form of prescience is operating in the evolution of any of these strategies for coping with future events. When organisms are able to adapt successfully to their futures it is either because by sheer chance they can do so, or because adaptations have been evolved in the past that may be able to cope with events in the future. Nature cannot know the future. At best it can 'guess' at what it will be like on the basis of past events.

A RÉSUMÉ AND A BRIEF NOTE ON TERMINOLOGY

I am telling a complicated story in as simple a way as possible. For this reason it is time to take stock of where we are, and to review the way words are being used. This is the story so far. We human beings seem to be rather good at knowing about the world. By knowing I mean being able to establish certain mental states that relate to certain features of the world. This fit between internal and external structure is knowledge. Now, what we know is, in an important sense, what we are as people. Yet our ability to know is not a purely human characteristic. Even if we confine knowledge to something that we have in our heads, then certain other creatures have it as well. However, we should consider knowledge, in the sense that it comprises that relationship between internal organization and external order, to be a particular kind of adaptation because the latter is identically defined. In that case, knowledge has a much wider significance. It is a deeply rooted characteristic of all organisms, and the way in which adaptations evolve should then be seen as a way of gaining knowledge.

If all adaptations are a form of knowledge, then, of course, so too are adaptive behaviours forms of knowledge. I am giving the conventional label of instincts to adaptive behaviours that are constructed by complex developmental processes from genetic instructions. Instincts are efficient and economical forms of adaptive behaviour, and most animals get by with instincts alone. Instincts, however, like all adaptations, have one potentially grave drawback: they are constructed on the basis of instructions built up in the past. Because nature is never prescient, this is all that they can ever be built on. But in a world that is always changing, such instructions may not be entirely appropriate for living in the world as it is now. Furthermore, certain kinds of change are so rapid that the processes by which adaptive knowledge is gained cannot detect such change. This is because such knowledge is gained by a process, which we have called the primary heuristic, that operates at a pace determined by how long it takes an organism to get from conception to when it itself is able to reproduce. In the event that such high-frequency change is significant for an organism, then that species of organism has to evolve special organ systems that can operate at these higher frequencies and hence can track such change. The process by which such special organ systems work we labelled a secondary heuristic.

Thus it is that a special class of adaptations have evolved: these are adaptations that compensate for the shortcoming of all other adaptations. There are two kinds of these. One, the immune system, operates in the sphere of chemistry. The other comprises the intelligence mechanisms of the brain and operates in the sphere of the physical world of temporal and spatial relationships of events and objects.

Now let us get some words straight. A heuristic, remember, is something that leads to discovery and invention. The evolution of adaptations is the invention and discovery of how to organize living structures relative to particular features of the world. That is why I called the genetic-developmental processes by which such invention and discovery come about the *primary heuristic*. The processes of discovery and invention making up the adaptations that compensate for the shortcomings of those formed by the primary heuristic I called the *secondary heuristic*. Those secondary heuristics that are concerned with tracking the spatial and temporal relationships of events and objects in the world are what I am collectively referring to as *intelligence*.

Learning and memory are forms of intelligence common to human beings and quite a wide range of other animals. The ability to manipulate stored knowledge – that is, reasoning and thought – is a form of intelligence most probably restricted to a small number of species. *Homo sapiens* is especially adept at this form of intelligence. We are also particularly good at sharing learned knowledge, which introduces a special dynamic into human intelligence. These are issues that will be discussed in the next chapter.

In summary, then, intelligence is the collective term for the secondary heuristic operating in the brain; learning and memory are its most common manifestations in different animals; reasoning and thinking are much more restricted forms of intelligence; and so too is culture, the ability to share knowledge, which adds another dimension to intelligence.

LEARNING AND INTELLIGENCE IN ANIMALS OTHER THAN HUMANS

There are three main groups of multicellular organisms. Two of these are the plants and fungi. Neither has the equivalent of a nervous system, and, of course, their members do not have actively to move through space in order to earn their living. Plants manufacture their own food, and fungi are parasitic on nutritional sources of other organisms, be they alive or dead. Any movement through space of members of either group is passive. They do not have locomotor organs and so do not have to employ internal energy sources to power such organs, as animals do. Subject to the same uncertain futures problem as all other living forms, they combat it with extraordinary reproductive cycles, in the case of fungi, and very extensive developmental flexibility, particularly in plants. Every gardener knows that plants possess astonishing plasticity of growth through all of their life spans. Such developmental plasticity is itself a form of tracking of environmental conditions and is undoubtedly the principal answer that plants and fungi have to the uncertain futures problem. There is absolutely no evidence of a learning and memory ability in either plants or fungi.

The other main group of multicellular organisms is the Animalia, which is divided by systematists into about twenty-five major groups called phyla. In five of these there is good evidence of learning. These phyla are the flatworms (for example planaria), the segmented worms (like earthworms), the molluscs (such as squids and octopods), the arthropods (which include insects, crabs and spiders) and the chordates (including animals with backbones, like ourselves). As can be seen, the insects, of which there are literally millions of species, are part of one of these phyla. As pointed out right at the beginning of this chapter, it is not known how many of these insects can learn in a way that is clearly differentiated from variable development, but it seems unlikely to me that many of them can.

This rather impressionistic view says something about the poor status of this area of science and also reflects a much wider problem than just our ignorance of insects. In how many species of fish, for example, can individuals learn? We know with certainty that a few can, because those are the ones that have been tested. I don't know of any study that has shown that the individuals of some fish species are entirely incapable of learning. But there are thousands of species of fish, and only the tiniest number of them have been tested. All that one can do is revert to guessing again, and in this instance my guess would be the opposite of that for insects. In other words, I would guess that most can learn, but we just don't know for certain.

In a larger context, we also don't know how correct is the intuition that learning is widespread among chordate species and rather more restricted among invertebrates, but if it is correct, then intelligence is an option for solving the uncertain futures problem that only a minority of living creatures have adopted. It is significant that the phyla containing the known intelligent species are mostly those characterized by vigorous activity. Animals that move are subject to self-induced high-frequency change, and I believe that such change has been an essential selection pressure in the evolution of intelligence.

Despite intelligence probably being a minority characteristic, it has been recorded and studied in hundreds of different species. (Knowing that the total number of extant species on earth is variously estimated as being between 10 and 30 million helps put this figure into perspective.) This is no place for an exhaustive review, but it is certainly worth pointing out some well-known examples drawn from entirely unrelated animals. The turbellaria are free-living flatworms with a position in phylogeny (which is the study of how species are related to one another) that is considered crucial by systematists. Thought to have important affinities with almost all the major phyla, the turbellaria were one of the earliest forms of animal to be bilaterally symmetrical (that is, having a single plane of symmetry, resulting in the two sides of the body being mirror images of one another); they also have a consistent front end during movement, in which are concentrated sensory organs and a mass of nervous tissue comprising a brain of sorts. So whether turbellaria can learn or not is of considerable interest.

A well-known and common kind of free-living flatworm is the planaria. One of the most convincing demonstrations of learning in planaria was carried out in the 1920s, but has since been replicated and improved upon several times. Such experiments have shown that planaria can learn, and remember over a period as long as twenty to thirty days, that if they swim along a wire, food will be found at the end. In effect, planaria learn to swim extremely simple mazes. Somewhat more controversial are claims that planaria can learn simple associations in a manner strongly resembling Pavlovian conditioning, and that such learning can be passed on to an untrained animal when the latter eats the former – literally a consumption of knowledge.

Segmented worms have a much more pronounced head end and brain, a brain being a concentration of nervous tissue which serves as a centre for co-ordinating signals coming from numerous senses with outputs to motor organs. Such worms have been shown to be able to learn a rather larger range of tasks than can flatworms. They show a definite form of Pavlovian conditioning, even if with some curious characteristics; and they are quite adept at learning their way through simple mazes, though the period of time for which such learning can be retained has been differently reported and remains unclear. Earthworms can also learn to avoid negative events as well as to approach positive events.

Honey bees, as already said, are miraculous little animals. They have exquisite behavioural adaptations by which they can navigate and communicate both the position and excellence of potential new nesting sites and the richness of food sources. So much attention has been paid to the famous dance by which they impart such information that the obvious fact that they are able to fix spatial positions in their memory is frequently overlooked. What is more, they seem able to retain a number of such spatial fixes simultaneously and to manipulate the ranking of the quality of resource that attaches to each. This is the only way to explain the exchange of information within a colony as to the excellence of alternative sites for a new hive and the way in which, eventually, a consensus is reached on where that new site is to be. Honey bees have also been shown to exhibit conditioning phenomena that are similar to those reported in mammals.

Birds are animals as different from bees as bees are from worms. Pigeons have become standard laboratory animals for the study of learning; but other species of bird have been studied as well, none more intensively than songbirds. Of particular interest is the problem of how they come to learn certain features of their song and, most intriguing of all, regional dialects. There are species like the chaffinch in Europe and the white-crowned sparrow in North America whose songs show systematic changes across a spatial range. These dialect differences are learned: a Somerset-born chaffinch raised in Sussex will sing with a Sussex dialect, and vice versa. This perceptual learning occurs through hearing the songs of local adult males when the birds are young and not yet themselves able to sing. Indeed, the whole behaviour of singing in a species like the chaffinch is heavily dependent upon several forms of early learning, including a form of motor-skill learning where the birds have to be able to hear their own vocal output if they are not to end up with a hopelessly distorted and malfunctional form of chaffinch song. There is also some evidence that during such learning songbirds are most sensitive to the effects of hearing their own species' song rather than the song of other, even closely related, species.

These findings are especially intriguing and tell us that birds like the chaffinch *know what it is they have to learn*. This may seem paradoxical, but it is easy to understand if you think of the brains of these birds being primed by the primary heuristic to learn only certain things and not others. We will return to this point in the next section.

The four examples that I have given of animal learning, intriguing though they are, are not typical of the great majority of experiments conducted on animal learning in laboratories over the past ninety years. These have concentrated on two species, rats and pigeons, and a restricted set of experimental situations which are descendants of Pavlov's original conditioning studies and E.L. Thorndike's puzzle-box experiments. Conditioning studies are based on the assumption that the learner can acquire associations between events in the world by those events impinging upon the learner's receptors under certain conditions – determining what these conditions are is the objective of such research. In conditioning the learner is a passive observer – it looks, hears and feels, but it doesn't have to do anything. Thorndikean experiments, by contrast, variously labelled instrumental, trial-and-error or operant learning, assess how behaviour is altered by learning about doing, where the animal acts on the world and those actions have consequences for the learner. So, while Pavlov restricted the movements of his animals and had them passively observing, say, the ticking of a metronome and the squirting of meat powder into their own mouths (subsequently the metronome alone leads to salivation, showing that a link, an association, has been forged between the stimuli), Thorndike had his cats actively effecting their own escape from the puzzle box, which then had the consequence of allowing them access to the morsel of fish placed outside of the box.

Such work has been carried out with great scientific rigour and, like all good science, has been based on the assumption that there is a limited number of laws of learning that await discovery, these limited laws finding expression in at least a degree of similarity of learning across different species. And for a long time the evidence for such generality seemed quite strong. A famous example concerns what are known as the schedules of reinforcement. If a rat or pigeon manipulates a lever or key, receiving a reward each time that it does so, and if suddenly the rewards are withdrawn, then the animal will quite quickly cease to contact the lever or key. But if the rat or pigeon has been rewarded more intermittently and not every time that it contacts lever or key, then when the reward is completely withdrawn the animal persists in the learned behaviour longer than if the reward had been consistently received after every response. This seemingly paradoxical effect, known as one of the partial reinforcement effects, is common to both rats and pigeons, and is widely assumed to occur in all animals capable of instrumental learning.

But problems do seem to begin when one tries to establish such effects in an animal like the Syrian hamster. Hamsters do not show the same partial reinforcement effects, and the most likely reason for this is that hamsters are hoarders. They don't consume rewards immediately. They hoard food, and their hoarding is an adaptive behaviour, an instinct, and in this case an instinct that seems to get in the way of establishing partial reinforcement effects. Because they don't immediately consume rewards, this means either that they are, in effect, always on partial reinforcement schedules, whatever the experimenter thinks they are on; or, that they are always continuously rewarded, because what hamsters really like is tucking food away for future consumption, the actual eating being less rewarding than the tucking away. Either way, partial reinforcement in the hamster leads to different characteristics of learned behaviours than occur in rats or pigeons learning under identical conditions.

Perhaps because learning theorists by chance chose to use as experimental animals individuals from two species that just happen to learn in the same way, or perhaps because there are real similarities between all learners, for much of this century data were gathered that supported the assumption of the existence of general laws of learning – such generality implying, it was assumed, that such laws applied to humans as well. However, the ethologists were always sceptical of the 'general process' approach to learning, because when they looked at animals living under natural conditions they saw different species as characterized by different behaviours. Now, this does not mean that there cannot be the same processes underlying the learning of such different behaviours, but then neither do consistent findings from just two species of learner prove that all species learn in the same way.

By the 1970s learning theory was in a troubled state. Actually it was, and remains, a state common to many areas of biology and it derives from a peculiar feature of evolutionary theory. Evolutionary theory says that the same evolutionary forces move and shape all living things. It also says that through the multiplicity of chance events that are built into the process of evolution, and which are compounded across history, living things will also display great diversity. So it is a theory that demands similarities and differences at one and the same time. It also never rules out the possible existence of common processes or features, which either derive from some common ancestral event and cannot easily be reversed, or consistently win out in competition against other processes or features because none are better. And to make things more difficult, we do know that some features of organisms, like the way in which genetic instructions are coded in DNA, are literally universal. So the trick is to arrive at an understanding that accommodates both universality and diversity simultaneously. This is a requirement for the science of intelligence in general, and learning in particular, just as much as it is a requirement in other areas of biology. The theory of the evolution of intelligence presented here provides that understanding.

THE NON-*TABULA-RASA* NATURE OF INTELLIGENCE

Intelligence is part of the secondary heuristic that evolved because of the temporal sampling limitations of the primary heuristic. Because intelligence can only operate as a device for dealing with predictable unpredictability, where the unpredictable element is a rather extended wobble (a short-term stability) on the predictable element, this means that the secondary heuristic is functionally always tucked under the wing, so to speak, of the primary heuristic. And this means that it is directed by the primary heuristic – the primary heuristic tells it, roughly, what to do. There is a technical phrase for this condition, which is that the two heuristics are in a relationship towards one another known as a nested hierarchy. Specifically, it is a control hierarchy where the scaling factor is the frequency of change in the world to which each is sensitive, and the primary heuristic is the more fundamental level.

Consider a simple example. It is clear that animals, including ourselves, are disposed to associate certain things more readily than others. The classic case is how easy it is for us to associate a feeling of nausea with tastes and smells, and how unlikely we are to learn the association between a feeling of illness and a place. This occurs in certain other species as well, including rats, the phenomenon being well known to pest-control experts who have labelled it bait-shyness. If animals like rats or mice survive poisoning, they will readily return to the site of the poisoning, but will not take again into their mouths and swallow anything that has the taste of the poison.

How can this be explained? To answer this we have to break the problem down into two parts. First there is the wider and more enduring association between nausea and taste in an animal like the rat or ourselves. We experience nausea when we have been poisoned; poisoning follows the ingestion of food; and food is recognized by its taste and smell, not by where we are when we eat it. This linkage between nausea and taste is the predictable component, which, if it changes at all, does so very slowly indeed. The primary heuristic, therefore, evolves brain states – adaptations

– that predispose towards the illness–taste association rather than the illness–place association. The second part of the problem concerns the unpredictable component, which is exactly which foods with what tastes actually lead to poisoning. This is something that may vary quite rapidly, in fact too rapidly, and so be too fine-grained a change for the primary heuristic to be able to detect. This is done by the secondary heuristic in the form of a simple component of intelligence that we call conditioning. This particular form of conditioning is primed by the primary heuristic to associate tastes and smells with illness – but not places, sounds and temperatures with illness. In other words, the primary heuristic, itself a knowledge-gaining process, 'tells' the secondary heuristic, also a knowledge-gaining process, what to learn.

In exactly the same way we can explain how songbirds learn from certain sounds only, rather than any sounds. The primary heuristic constructs a songbird brain that has a template specifying the range of sounds to be learned. Within that range is the actual song learned by the secondary heuristic of song learning. That is how the birds know what it is they have to learn.

Lorenz, though unable to provide a proper conceptual scheme to explain why, knew that animals learn only certain things. No learner is a generalist able to acquire knowledge about any and every perturbation in the world that occurs within the frequency band of change to which the intelligence of any one species is tuned. And he invented a lovely phrase as an explanation for it. All learners, he claimed, have in their brains what he called 'innate teaching mechanisms' that lead to the learning of only limited classes of all possible learnable events. These innate teaching mechanisms are what I have been referring to as the adaptive brain states furnished by the primary heuristic to prime and direct the activities of the secondary heuristic. The conception of a constrained intelligence, as opposed to a generalist, free-wheeling intelligence, is perhaps the most important notion that comes out of evolutionary epistemology.

This nested hierarchy of knowledge-gaining processes has two strong implications that are important for philosophy and psychology. The first, an extension of what has just been said, is that John

Locke's notion of the mind as a *tabula rasa*, a blank slate on which experience – any experience – is written, is quite simply wrong. All intelligence is limited by the primary heuristic. The slate is already written on when individual intelligence begins to operate, and all that it can add to the slate are the fine details of the themes already written there. Actually, the case can be stated even more strongly. This *has* to be so, because if it were otherwise, intelligence would not exist, that is, it would not have evolved at all. This is because, if intelligence operated on a kind of 'blank slate principle', it could not function at the speed that it does and *must* function at – that speed reflecting the frequencies of change to which it is sensitive. That, after all, is why it evolved in the first place.

If learners had to search through every energy change that occurs in their environment and act on it to test its consequences before deciding what to learn, even if they had the sensory equipment to do so, which they do not, then learning would be an impossibly slow process. But learning is not slow. It is a remarkably rapid way of gaining knowledge. That can only mean that learning does not ever start from scratch, but starts from a position at which the learning mechanisms know what it is that they have to learn. It is also the case, speed apart, that if learning did operate in this undirected way, that it would most often result in behaviours that are maladaptive. However, learning almost always results in adaptive behavioural change – another of Lorenz's great insights. This too can only mean that learners start off knowing what has to be learned.

So the primary heuristic constrains what the secondary heuristic can do. Now, we know from Chapters 2 and 3 that the primary heuristic gains its knowledge a posteriori. The constraining of the secondary heuristic means that the primary heuristic feeds to it knowledge already gained at that more fundamental level in this hierarchy of knowledge processes. That is why the secondary heuristic, intelligence, appears to have knowledge a priori; it does indeed have a priori knowledge, but it is knowledge that was initially acquired a posteriori by the primary heuristic. The nested hierarchy thus saves us from having to invoke any contra-causal explanation of such a priori knowledge.

The second important implication is that it resolves the old nature–nurture problem that goes back at least to the writings of Plato in a manner different from the currently widely accepted resolution. This problem has come to us down the millennia in a number of different forms and warrants a book to itself. Most recently, there have been claims that it has been solved, these claims going, in general, as follows. 'The instinct–intelligence and genes–experience dichotomies, or any other of the dichotomies that characterize the nature–nurture distinction, are false dichotomies. This is because all genes require an environment in which to develop. Since development requires both genes and experience, and since every trait is constructed by a developmental process, every trait requires both nature and nurture.' End of problem.

In my view, this is a fundamentally wrong approach to the matter. The solution that it provides is 'horizontal' in that it still maintains the separation of genes, coming as it were from one side, and the environment, coming from the other, with some kind of developmental integration in the middle. It is called interactionism, with the interaction purporting to get rid of the separateness of the two components. However, the nested hierarchy scheme outlined in this chapter is a quite different kind of resolution, because the scheme is 'vertical'. Intelligence, the secondary heuristic, is subsumed under, enclosed by, the primary heuristic. The horizontal components of interactionism, the internal and external causes, remain present at both levels. They are not the issue. The issue is some superordinate concept that subsumes all levels of the hierarchy, and that concept, I am arguing, is adaptation: knowledge. Intelligence nested under development and the genetic process does not allow any claims about such-and-such being caused either only by nature or only by nurture or by both. Such language and imagery are wrong. Intelligence *is* an adaptation, and the required integration can only be achieved vertically, not horizontally.

This vertical integration of the primary and secondary heuristics is the refutation of what in Chapter 4 I referred to as the doctrine of separate determination. The latter, remember, is the long-held view that behaviours that are controlled by rationality and intelligence (the secondary heuristic) are to be viewed as quite separate from those controlled by instinct (the products of the primary heuristic). But the secondary heuristic is directly and closely related to the primary heuristic. The secondary heuristic has its origins in the temporal sampling limitations of the primary heuristic and functions to make good this limitation. Rationality and intelligence are extensions of instinct and can never be separated from it. The doctrine of separate determination is completely wrong. As with so-called 'interactionism', the imagery and the concept of separate determination are false. *Instinct is the mother of intelligence*.

There is one final point to be made here within this general context. It is that intelligence is not some functionally unitary entity. Conceptually it is all one, all a part of the secondary heuristic. But because of its nesting under the primary heuristic, intelligence cannot easily and automatically be equated across species. Intelligence in the honey bee is a set of knowledge-gaining processes whose specific function is set by the primary heuristic of the honey bee. This means that honey bee intelligence is intimately linked to honey bee genes. Precisely the same argument applies to pigeons, rats and humans. Rat intelligence must be understood in the context of rat genes, and human intelligence can only be understood in the context of human genes. In so far as rat genes are different from human genes, then so too is rat intelligence different from human intelligence.

Species-typical intelligence rather than some identical intelligence that spills over from one species to the next is what I am arguing for. However, the notion of multiple intelligence*s* rather than some single intelligence does *not* imply that intelligence operates through necessarily different processes in different species. Quite the contrary: it is much more likely that the process of intelligence is usually the same across species, especially species within some restricted taxon such as a class (mammals or birds) or supraclass (vertebrates). What leads to multiple intelligences is the way the process interdigitates with other species-typical adaptations.

SELECTION AND INSTRUCTION REVISITED

What might this process be by which the secondary heuristic of intelligence gains knowledge? There are only two candidates, both already described in Chapter 2: selection and instruction, corresponding to Darwinian and Lamarckian theories of evolution respectively. The principal difference between them is that selectionism involves an over-proliferation of entities, the generation of which is unconnected with the organism's needs at the time (which is why the process is sometimes referred to as blind); a small number of them are conserved after they have been tested against the organism's requirements, and these are then mixed with other variants at the next stage of proliferation. In the case of intelligence, these entities are brain states that form the internal end of the knowledge relationship. Instructionalism, on the other hand, involves the production of an entity, a product of intelligence, after the need for it has arisen (which is why the process is considered to be directed); this process requires a malleable substrate that is moulded into an adaptive state by the environmental feature that forms the external end of the relationship. Put crudely, in the case of instruction the environment rules; in the case of selection, internal or organismic states lead.

Before considering arguments as to which, if not both, should be thought of as the process by which the secondary heuristic of intelligence works, several general points must be made. First, the end result certainly does not dictate the mechanisms by which it is reached. There are several ways, for example, that an infant might come to associate its mother's facial features with her smell (which is the first aspect of a mother that babies get to know, vision coming later). One way may indeed be a form of Pavlovian conditioning in which those brain states corresponding to the baby's representation of its mother's smell are linked with those brain states corresponding to the infant's representation of its mother's face, the association being 'stamped in' or 'impressed' by the constant conjunction of the two in the infant's world. Here the association is directed by the environment, and one would look for mechanisms reflecting such passive formation of knowledge.

Alternatively, the brain of the infant may internally generate multiple representations of the mother in which are mixed components of the actual mother as well as features of the world that are not the mother, and which may not even be habitually associated with her. These multiple representations of the mother are tested perceptually against the appearance of the mother on many occasions, as well as being acted upon to test the accuracy of the representation. The result is a complex brain state representing the mother in smell, facial features, sound of voice and even language characteristics – but eventually excluding those more transient features of the world that are, in fact, not the mother, such as certain clothes, places, food smells, and so on, that are regularly paired with the mother but are not actually a part of her. Thus it is that the infant *creates* a representation of its mother by generating, possibly sequentially and possibly simultaneously, numerous brain states representing her, and successively selecting out and reducing these brain states until only one composite brain state is left. This, of course, is a selectionist process.

In both cases we end up with the knowledge of mother. And in both cases the result might be a rich association of characteristics. But the mechanisms are very different.

Now, if the secondary heuristic does work by a selectional process, then, like the primary heuristic, it must have the components of the g-t-r heuristic described in Chapter 3. Remember also that one way in which the g-t-r heuristic can be characterized is by the mix of novel and already existing variants that the heuristic tests and then regenerates. A conservative mix will reduce the radical or chance component towards zero and keep generating and regenerating into the future a restricted number of variants. Certain forms of associative learning, like Pavlovian conditioning, are conservative in just this way. But the conservativeness is a result of a conservative setting of the g-t-r heuristic for that particular form of learning. What we have here is an instance of a selectional process mimicking instructional learning.

There are, of course, many 'kinds' of learning. Some do result from passive observation; others involve action on behalf of the learner; and yet others again require internal manipulations and computations of learned knowledge to result in higher-order knowledge forms such as response strategies and complex perceptual structures and skills. All could be served by selectional processes involving different variants and different settings for the ratio of radical to conservative variant generation. Both the nature of the variants and the settings of the ratios themselves are a part of the priming by the primary heuristic that results from the nested relationship between primary and secondary heuristics.

Complicating much of the debate over whether learning occurs by way of instructional or selectional processes and mechanisms is the potential for confusion that arises from using the word 'association'. The word is used sometimes in the sense of a causal mechanism, sometimes in the sense of a consequence of a mechanism that may or may not be associative (that is, instructional) in nature and sometimes in the sense of how the world 'out there' is to be understood and represented. With regard to the last of these, the temporal, spatial and causal texture of the world has, of course, to be described in terms of what connects with what. 'Mother' is a set of characteristics tied together in time and space. She is made up of certain intrinsic relationships (her eyes are blue, her skin is freckled) and extrinsic causal relationships (she is closely connected to the pleasurable sensations of food, kissing and cuddling). Mother is something out there whose characteristics become bound together into a coherent internal representation in which those associated external characteristics are mirrored by brain states of appropriate associations. This is the knowledge of mother, a mass of associations. But the associations that make up the representation of mother can be acquired by selectional mechanisms and processes.

Finally, one must concede the messy possibility that, unlike the primary heuristic which is wholly and solely selectional in form, intelligence might comprise both instructional and selectional processes. Indeed, the secondary heuristic may, in the course of individual development, be primed in part to engender instructional processes as part of a wide array of devices that make up the secondary brain heuristic in animals like ourselves and other vertebrates where that secondary heuristic has evolved into something that is complex and multiply formed.

The fact is that right now we do not know how the brain learns, thinks and does all the other things that go to make up intelligence – at least we do not know in anything like sufficient detail for one to be able to say with certainty whether it works in a selectional or an instructional manner. Traditionally, the technical neuroscience literature has been based primarily upon instructionalist theory. However, in the last twenty years or so, elegant and powerful theorizing by the likes of Jean Pierre Changeux in France and Gerald Edelman in the United States has shown that selectionist models of brain development and function are both plausible in terms of known mechanism and interesting in their psychological implications. My own preference, clearly, is for selection, and I shall explain why shortly.

If I am correct, then another line of reasoning must follow. If the brain is a 'Darwin machine' in the sense that at least part of the brain's secondary heuristic works as a selectional process, or a set of selectional processes, then we must ultimately be able to describe and understand the way that intelligence works in terms of the generation of variants and the machinery for internally selecting from these variants a restricted set that must be conserved and hence propagated (transmitted) into the future. We should also be able to use that second descriptive framework of universal Darwinism, the replicator–interactor–lineage conception given in Chapter 3. Remember, according to this set of ideas, all Darwinian evolution can be understood in terms of the existence of replicators, which are entities able to make copies of themselves; interactors, which are entities that interact with the environment to result in differential replicator selection; and lineages, which are the entities that change in time as a result of replication and interaction. What, in terms of behaviour and its accompanying brain states, might correspond to the replicators, interactors and lineages of the secondary heuristic?

A replicator is a pattern of activity in a nerve sheet, the settings of the connections between nerve cells in the jargon of current neural net modelling, and corresponds to a memory. Memories are multifaceted things. Each memory is made up of different combinations of activity in many nerve cells, and some memories share overlapping patterns of cells and activities of cells. Changes, often small changes, in the cells and their patterning of activity will lead to changes of memory, and memories may thus grade into one another. The memory of mother smiling involves a somewhat different set of cells from the memory of mother scolding. Because the nervous system functions in a way that is much more fluid than the way that the genome of each cell functions, memories may, and do, constantly slip and slide into other memories. But the memory of mother smiling is a relatively fixed pattern of activity in a relatively fixed population of brain cells. This means that given the fluid nature of brain function, and the enormous information load that is placed on the brain, the pattern of activity that constitutes the memory of mother smiling is not always present – it is usually a potential pattern rather than one that is actually present.

This is just another way of saying what we all know to be the case, that we do not live with all of our memories activated all of the time. But they can be re-constituted, re-membered, re-plicated. (The word 'replicate' has its root in *plicare*, meaning to fold. Add the prefix and you have the word meaning to form again by folding – good imagery for genetic replication, if a bit less so for brain cell pattern replication.) Memories *are* replicators. They are the secondary brain heuristic analogues of genes. They do not have the mystical force attaching to them in the secondary heuristic that Richard Dawkins ascribes to the DNA, the replicators of the primary heuristic. That may be because they are nested under the replicators of the primary heuristic – intelligence being linked to and in the service of the genes. Or it may be that replicator mysticism is simply misplaced. Either way, there is no difficulty in thinking of memories as replicators in the sense that memories are produced again and again as copies each time that the brain state in question is reconstituted – that is, each time we conjure up some specific memory.

The knowledge embodied in the replicators of the secondary heuristic, like all knowledge, usually must be good working knowledge in the sense that it must relate to some specific feature of the external world and with sufficient accuracy (not perfect, just sufficient in H. A. Simon's sense of satisficing) that it can be acted upon and the outcome will be adaptive behaviour. Such replicator-determined actions, adaptive behaviours powered by memory, are the interactor components. Interactors, remember, are entities that lead to differential replication. If the child's memory of its mother smiling is acted upon, and the resultant behaviour is inappropriate and punished in some way, then the memory of the mother is altered and replicated as a more complex memory. Sometimes the mother doesn't smile, and the causal texture within which mother smiling and mother scolding occurs is then stored and further acted upon, as either a single complex memory or a set of related memories. These actions, the interactors, are tested by their consequences, and when finally the whole complex of mother replicators is in place (assuming that we can arbitrarily stop the whole process at some moment), then we will have a lineage of the representation of the mother in the endlessly changing brain of the child. These changes, remember, are the outcome of an intelligence specifically evolved by the too-slow-acting primary heuristic to track the rapidly changing features of the world like mother and the circumstances under which she smiles or scolds.

Returning now to the question raised at the start of this section, why should the brain be seen as a Darwinian kind of machine rather than as a Lamarckian machine? There are three answers. The first is that you have to choose one or the other to formulate a hypothesis which can then be tested. Put rather more crudely, you do not have a problem in science unless you think you have an answer; unless you come down on one side or the other, for selection or instruction, then you don't have a coherent formulation to think about and test either conceptually or empirically. (In the next chapter we will briefly consider science itself as a part of universal Darwinism.) You can't fence-sit in science. For science to work, people have to take sides so that ideas can be tested against the world. A shapeless intellectual structure made up of six of one and half a dozen of the other is bad science because it will predict anything and so cannot be properly tested.

Forced to take sides, then, there are two further reasons for choosing the selectionist camp. One is the problem of creativity. Intelligence, at least in humans, does not involve just the faithful tracking of changing events. It involves also the production of novel solutions to the problems posed by change – solutions that are not directly given in the experienced world. What such novelty means is the contribution to that intelligence, to knowledge, by factors internal to the knower and hence unique to that individual intelligence. Such creativity cannot occur if change is slavishly tracked by instructionalist devices. So what we see here is that while selection can mimic instruction, the reverse is never true. Instructional processes can never lead to creativity. Instructional intelligence comprises only what has been actually experienced. To go beyond experience requires the generation of something from inside the knower, and only an intelligence driven by selectional machinery can do that. Indeed, according to D.T. Campbell, the father of modern evolutionary epistemology, selectional processes are required for the acquisition of any truly new knowledge about the world: 'In going beyond what is already known, one cannot but go blindly. If one goes wisely, this indicates already achieved wisdom of some general sort.' Instruction is never blind. Selection always has an element, at the very least, of blindness in it. At the heart of all creative intelligence is a selectional process, no matter how many instructional processes are built on top of it.

The third reason for choosing selection over instruction is one of parsimony and simplicity. If the primary heuristic works by selectional processes, which it most certainly does; if, as will be argued in the next chapter, culture works by selectional processes, which is fairly widely agreed to be the case; and if that other embodiment of the secondary heuristic that deals with our uncertain chemical futures, namely the immune system, works by selectional processes, which is now universally agreed: then why should one be so perverse as to back a different horse when it comes to intelligence?

A nested hierarchy of selectional processes is a simple and

elegant conception of the nature of knowledge. There will have to be good empirical reasons for abandoning it.

INTELLIGENCE, REDUCTIONISM AND THE CAUSAL SHAPE OF EVOLUTIONARY THEORY

Some years ago United States space scientists put an unmanned vehicle on Mars. It was called Mars Rover and its job was to carry out certain observations and experiments on the surface of Mars. But Mars Rover presented a problem to its controllers. At that time the distance between the Earth and Mars was such that radio signals sent from Mars to Earth and back took more than eight minutes. If the vehicle's behaviour were controlled only from Earth, and if as Mars Rover moved across the surface of Mars its cameras were to send a signal back to Earth showing a large green figure with seventeen eyes on stalks suddenly appearing in front of it, or more realistically that a hole lay before it, then by the time that signal had reached Earth and the controllers had sent a return message telling Mars Rover to back away making placatory noises, or to turn, Mars Rover would have been vaporized by an unfriendly Martian or ended upside-down at the bottom of a Martian hole. Mars Rover would not have survived. The transmission times meant that special provision had to be made by which a Mars Rover vehicle could survive and fend for itself.

The engineers building Mars Rover had several choices. One was to send lots and lots of vehicles to Mars, not just one, each hard-wired with slightly different instructions (for example, one might combine the instruction that when confronted with a purple figure with sixty-four eyes it should back away silently, turning at a certain angle should the ground in front of it begin to slope downwards at an angle greater than 45°; another might define a Martian as having different characteristics and a Martian hole as involving a downward angle in ground surface greater than 60°; and so on) because those engineers had no way of knowing exactly what hazards a Mars Rover vehicle might run into. They might also have given it faster work time and travelling speeds so that it could complete the required experiments in a shorter period – a kind of Mars Rover r-strategy. A second possibility was to send a Mars Rover bristling with armour, armaments and devices for staying upright – that is, giving it very specialized structural adaptations rather than avoidance behaviours.

A third choice was to build a limited artificial intelligence into Mars Rover, and this is what they decided to do. Into Mars Rover's 'brain' was built a set of simple rules by which it could order its experimental programme according to local conditions (for example, if on a rocky surface, don't try to do the soil experiment, measure solar radiation instead); and also a small learning capacity constructed around a limited number of concepts which, in effect, defined the limiting conditions signalling possible hazards (for example, a Martian moves under its own power) and allowing Mars Rover to fill in the details (what colour it is and how many eyes it has). Equipped with this simple intelligence, Mars Rover was able to survive longer and to operate more effectively, because it was able to order its behaviour according to the conditions of Mars as it actually encountered them.

There are considerable parallels with intelligence in animals. The lengthy transmission times between Mars and Earth, which cut off Mars Rover from its Earthling controllers, are similar to the cutting off of an animal from the instructions in its gene pool after conception. The unknown hazards on Mars are Mars Rover's uncertain futures problem. The engineers' plans and Mars Rover's construction here on Earth are equivalent to the primary heuristic. Mars Rover's highly constrained artificial intelligence is the equivalent of a secondary heuristic, and one, furthermore, which is most certainly *not* a *tabula rasa* learning system.

What is most instructive about Mars Rover is the way it makes us think about the causes of its behaviour. Some days after arriving on the surface of Mars it was adept at avoiding holes and breezily skirting around Martians. How can this be explained causally? One cannot just point to the plans for its construction here on Earth; such blueprints and the subsequent building of Mars Rover clearly are part of the cause of that behaviour, but not all of it. The other part lies in the actual operation of the artificial intelligence in Mars Rover's head. What the engineers had done was give Mars Rover *the ability to generate its own causes for its behaviour*. This, after all, was the reason they had to build an artificial intelligence capacity into it in the first place. The engineers here on Earth were unable to predict with certainty everything that Mars Rover might encounter on Mars and so were unable themselves to provide Mars Rover with those causes. Indeed, had they been able to do so, then they would simply have hard-wired Mars Rover in the appropriate way. But not being able to do more than predict roughly what the hazards might be, they left it up to Mars Rover to supply its own causes for its behaviour by filling in the gaps in the predictable unpredictability formulation. So Mars Rover had to gain some of its knowledge by its own intelligence and, by testing how to respond to and interact with these features of its environment, build that knowledge up into working knowledge. A causal explanation of Mars Rover's behaviour, therefore, has to take into account both the engineers' specifications here on Earth, equivalent to the genes of intelligent animals, *and* the operation of Mars Rover's artificial intelligence when on Mars, equivalent to the secondary heuristic of intelligent living creatures.

It is worth pushing the analogy just a bit further. If Mars Rovers had been sent in quantity to Mars, each different in their engineering specifications and behaviour, and if they were destroyed or wore out after a while, but if those Mars Rovers that could survive the Martian hazards could replace themselves with new Mars Rovers by exchanging their engineering specifications to give rise to new vehicles which combined the attributes of their 'parents', then – given that some small percentage of the vehicles were 'intelligent' and hence better at surviving than those that were not *because they could generate the causes of their own behaviour* – the population of Mars Rovers would slowly change to one dominated by smart Mars Rovers. The population would have been transformed in time; that is, it would have evolved; and one of the causes of that evolution would have been their intelligence.

Now, the same logic applies to a causal explanation of the

behaviour of any living thing that is intelligent, and to the causal role of that intelligence in the evolution of the populations to which such intelligent animals belong. The causes of intelligent behaviour are divided between the genes and development (the primary heuristic), on the one hand, and the capacity to gain knowledge and act on it through the operation of the secondary heuristic, on the other.

Furthermore, it is only if the secondary heuristic works by way of a selectional or Darwinian process that the secondary heuristic, intelligence, is able to generate the causes of its own behaviour. If the secondary heuristic operates exclusively by way of instructional processes, then the causes of intelligent behaviour would be entirely reducible to genetics, because intelligence would merely passively reflect the nature of those instructional processes and the circumstances of the world that are imposed upon it. Intelligence would become merely a rather extended but entirely deterministic device understandable entirely in terms of genetic determinism on the one hand and the history of organism–environment interaction on the other. What saves intelligent behaviour from such a reductionistic account is the presence of selectional processes in the mechanism of intelligence. As long as the secondary heuristic operates, even if in only small part, by Darwinian processes involving the unpredictable generation of variants, then the products of that secondary heuristic, intelligent behaviour, cannot be reductively explained by genetics or genetics and development.

The converse of the Mars Rover argument also applies. Animals that have not evolved a secondary heuristic that operates on Darwinian processes, be it a heuristic for dealing with either the chemical or the physical uncertain futures problem, can be explained entirely by a reductionist genetical account. No matter how complicated the interactions between organisms and their environments, no matter what co-evolutionary mechanisms exist by which organisms are coupled to each other and to their environments, a genetic reductionist account of the causes of such evolution will be sufficient, and satisfyingly parsimonious. It is only when the evolutionary picture is added to vertically, with nested Darwinian processes tied into a control hierarchical structure, that the causal shape of evolutionary theory has to be stretched beyond the genetic-developmental level to encompass other units of selection – that is, other replicators, interactors and lineages – than those traditionally evoked by conventional evolutionary theory. It is only then that genetic reductionist accounts of evolution fail.

Genetic reductionism can never be invoked, then, as an explanation of the behaviour and evolution of animals that are intelligent. That, of course, includes *Homo sapiens*. And this argument allows us to return to the point made by the molecular biologist John H. Campbell quoted in Chapter 2 about how events are directed, caused and carried out during evolution. Remember that the gist of his argument is that recent discoveries of molecular biological mechanisms have saved evolutionary theory from the notion of passive organisms being dominated by external selectional forces. The argument being advanced here is that equally important in redressing our understanding of evolutionary causes is the need to consider what intelligent animals have inside their heads. Once intelligence has evolved in a species, then thereafter brains have a causal force equal to that of genes.

There is another quotation from Chapter 2 that should be remembered in this context. This is S.J. Gould's invocation of hierarchy theory as the appropriate architecture for the theoretical complexity of an adequate evolutionary theory to replace the minimalism of neo-Darwinian uni-level theory. What Gould and others have claimed is that neo-Darwinism just isn't complete enough to qualify as the central theorem of biology. If evolutionary theory is to tell us about ecology, about ontogeny and about speciation, as well as about individual organism adaptation, then it must be expanded into some hierarchical form with more than one level and with more than one unit of selection. Gould is correct. But he left individual intelligence and behaviour out of his account. If evolutionary theory is to be complete, then intelligent behaviour has to be included, and so too does a very peculiar feature of intelligent behaviour in our species, namely culture.

I am no more certain than anyone else as to what the shape of future evolutionary theory will be. What does seem to me to be likely is that the shape will be significantly determined by the way we come to understand better just which are the irreducible hierarchies in nature, and how they are linked to one another.

SUGGESTED READING

As in the case of universal Darwinism (Chapter 3), there are no elementary or easily accessible accounts of the evolution of learning and intelligence. The following are some of the writings that have been inspirational for me. They are not easy to understand.

Lorenz, K.Z. (1969) 'Innate bases of learning', in K. Pribram (ed.), *On the Biology of Learning*, pp. 13–91, New York: Harcourt. Here Lorenz came much closer to teasing out the relationship of learning to instinct than any of his critics were ever able to do.

Sommerhoff, G. (1950) *Analytical Biology*, Oxford: Oxford University Press. A neglected classic that considers, among other things, the problem of time-lags in adapting systems.

Waddington, C.H. (1975) *The Evolution of an Evolutionist*, Edinburgh: Edinburgh University Press. A collection of essays from someone who, though he wrote little directly on the evolution of learning and intelligence, greatly influenced the writings of others, notably Piaget.

Aspects of Human Knowledge

═══════

Modern evolutionary biology makes possible a science of knowledge. The study and understanding of knowledge – a knowledge of knowledge – are no longer only the province of philosophy. What follows is a consideration from the point of view of evolutionary biology of some areas of human cognition that relate to the evolution of intelligence described in Chapter 5. While all adaptations are a form of knowledge, in this chapter we will be concerned with what is commonly called human knowledge.

It takes little thought to realize that our ability to know is prodigious. No figure can be placed on our memory capacity, but the number of items of information that can be retained in memory is very, very large – indeed, it is so large that most psychologists no longer consider it profitable to think of memory as a unitary phenomenon. Motor skills like tying a bow or riding a bicycle are built upon memory, but upon a form of memory that has certain features which seem to make them different from our knowing what the product is of 6 and 7; and both are different from the spatial maps that we have of the layout of our homes, which in turn seem to be different from the rag-bag of 'facts' that we all carry about in our heads, and which seem to be 'functionless' knowledge.

Furthermore, our memory, as we shall see, is not always some static thing. Until neurological damage intervenes, 42 will always be remembered as the product of 6 and 7. But the memory of what we ate at our last Christmas meal may be less than complete and may even contain items that were not actually consumed that day. Also, because we can carry out certain operations upon our memories, which is to say because we can think, many of our memories are constantly combining and regrouping with other memories, and hence constantly altering what we think we know. In other words, we alter the meaning of our knowledge for ourselves. Add to all this the emotional colouring that accompanies both our memories and our thoughts, and what emerges is a picture of astonishing richness and complexity. We seem to know so much, and in such varied ways.

Human knowledge is a product of the same primary and secondary heuristics as that of any other intelligent creature. However complex and subtle human knowledge may be, its evolutionary origins are no different from that of the knowledge possessed by any other knower, and it must be assumed to have the same general characteristics. Furthermore, it is helpful to keep separate the rather simple notion of knowledge as a product of certain processes and mechanisms and the myriad processes and mechanisms themselves by which those matching relationships between internal organismic states and external features of the environment are established. This is not always an easy distinction to maintain, because the internal end of the relationship, the brain and psychological states, will be at least partly, and perhaps very significantly, determined by those processes and mechanisms, which may also influence the nature and scope of what can be matched to the outside world – language is the obvious case in point. Yet, as we will see, despite its complexities, language is one of the strongest sources of evidence available as to the extent to which human knowledge shares certain fundamental characteristics with the knowledge that other creatures have of their worlds.

So, while this chapter may, with some justification, be thought of as written with not a little impertinence, I do believe that its general tenor is correct. Two additional introductory comments are needed. First, what follows is *not* an attempt to solve the central issues of contemporary cognitive science. It is not even a review of what little is known – which, little as it is, is enough to fill at least one book of its own. Instead, this chapter is a consideration of human cognition from an evolutionary biologist's eye view – and only certain selected aspects of human cognition at that, these being those that relate best to the evolution of intelligence as described in Chapter 5. Second, it must still not be forgotten that what we commonly understand by the word 'knowledge' is closely related to what evolutionary biologists call adaptations. All adaptations are forms of knowledge. So a complete account of human knowledge in this wider sense would take in all human adaptations, including such diverse characteristics as bipedal gait, the form of our hands and specific features of our metabolic machinery. It would also have to include what is known of human instincts, that is, adaptive behaviours furnished only by the genetical and developmental processes of the primary heuristic. This is an undertaking far beyond the scope of this book. If the reader takes the general point of the fundamentally biological nature of knowledge, of its relationship to adaptations and of how all adaptations are forms of knowledge, he or she will have absorbed that message by now.

So in this chapter I am going to concern myself with what is commonly called human knowledge. The psychological states and brain states of the secondary heuristic will be considered quite specifically in the light of that nested relationship between the primary and secondary heuristics.

HUMAN INTELLIGENCE: A *TABULA RASA* OR NOT?

In Chapter 5 it was argued that learning and intelligence had evolved in order successfully to establish adaptive matching relationships between the learner (the intelligent creature) and certain of the short-term stabilities of the world. The secondary heuristic, because it is nested under the primary heuristic, is *always* primed in some way rapidly to gain particular forms of knowledge – that is, matching relationships to particular short-term stabilities – rather than being set ponderously to acquire knowledge about any and every short-term stability in the learner's world. If that argument is correct, then we should expect human knowledge, certainly in terms of its object or content and possibly in its form as well, to be constrained. We humans should learn about a relatively restricted set of features in our world; and this restriction may also be manifested in biases in the way we think and reason. What evidence is there to support these assertions?

At this point we must make a small excursion away from the main theme into the realms of the nature of evidence that may, or may not, be used for or against such an argument. Science should never be carried out according to the Baconian principle that what scientists should do is merely collect data, the 'facts', treating all data as equal, in the hope that eventually, when enough facts are gathered in, the truth about the nature of the world will somehow leap fully formed from the mass of observations. If it were, then, to quote Darwin, one 'might as well go into a gravel-pit and count the pebbles and describe the colours' as do anything else. But real understanding about the world has never been revealed by data alone. Science isn't like that. It doesn't treat all data, actual or potential, as equal. 'How odd it is that anyone should not see that all observation must be for or against some view if it is to be of any service,' to quote Darwin again. What Darwin was advocating is the deductive method in science by which theory is disciplined by observation and experiment.

Theory itself may range in possible form from a rather vague guess as to the nature of things through to some formalized calculus, some axiomatized deductive system of signs by which predictions are strictly derived from hypotheses. Whatever the form of the theory, there is a sense in which it is a guess at the nature of the world, and we then test the guess – what Popper calls conjectures and refutations. The truth comes from an interplay between the creative imaginings of scientists and the merciless, unyielding findings of empirical test.

Now, certain aspects of evolutionary biology, notably behavioural biology, present a lack of balance between theory and empirical test when the theory concerns events in the past that formed the selection pressures that resulted in (caused, if you like) some characteristics of present-day species. The theory might be interesting, and intuitively plausible, but how can it be tested? Take as an example of considerable pertinence to this book what has come to be known as the *social function of intellect* hypothesis put forward by a British psychologist, N.K. Humphrey, some years ago. This hypothesis states that 'higher' intellectual functions (by which Humphrey means reasoning and thought as opposed to learning and memory) evolved not for the gaining of simple factual knowledge about the world (for example, there is water in this place), not for the mastering of skills and methods for exploiting resources (using a stick to knock down fruit that is out of reach, for instance) and not even for learning such facts or skills by observing and imitating others, but rather in order 'to hold society together'. By this Humphrey meant that the social group is at once both a crucial feature in the survival of social animals, and also the source of that social life which presents the highest intellectual challenge that social animals have to face. Social life, the argument goes, involves the preservation of a group structure in the face of individual tendencies to exploit and manipulate others; individuals must calculate the consequences of their own behaviour and assess the likely behaviour of others on the basis of projecting themselves into the place of others – reasoning that 'I would do such-and-such in his or her situation, but he or she is different from me in this and that way, and therefore I can expect him or her to do the following, which I can influence by behaving like so or like so.'

Thus the centripetal forces caused by the advantages of group life are balanced by the centrifugal forces of the costs of group living, the whole dynamic structure being mediated by intellectually controlled social interactions and calculations.

The result, following aeons of selection, might be a few individuals with rare and rarified intellectual abilities that allow them to write poetry, compose music or invent extraordinary machines. Such rare talent, however, is epiphenomenal. No one believes that the intelligence required for the doing of mathematics or the writing of fine books arose because of the need for writing mathematics or books. The evolutionary origin of primate intelligence, including that of humans, is at least in part that of social transaction and social abstraction. Social exchange and contract within the social group have been a crucial selection force for the primate secondary heuristic. Humphrey argued that 'To do well for oneself while remaining within the terms of the social contract on which the fitness of the whole community ultimately depends' requires a high degree of intelligence or rationality. Furthermore, such social activity, such 'politicizing', takes time, and as social groups become more complex, so more time is required to service such activity, and this reduces the time available for interacting with the non-social world and extracting from it crucial resources. Hand in hand with increasing social complexity, therefore, came the need for improved technical and 'factual' mastery of the environment, because resources have still to be extracted from the world but in less available time. Thus it is that the engine of social interaction also drives advances in non-social intellectual skills.

This argument for the social function of intellect certainly sounds plausible, but plausibility alone is not good enough. It may even have some power in explaining seemingly refractory phenomena such as the tendency we have to 'bargain' with, and enter into contracts, with the inanimate, or to create abstract entities, God or gods, with whom proto-social relationships are established. In the language of Chapter 5, Humphrey's hypothesis is that the human secondary heuristic is primed by some kind of sensitivity to things social, and possibly to operate with some form of 'social logic'. We just can't help but think of inanimate objects or abstract forces in social terms. Such explanation of the seemingly inexplicable, however, is not enough either. It is one thing to say that social exchange – co-operation and competition – is the warp and weft of human intelligence, but how do we go about getting evidence for or against such a hypothesis? It is only evidence that allows us to maintain a hypothesis; and, apart from biochemistry, bones and a few tools and other artefacts, human evolution is unavailable to us.

The nature of early human social groups, the numbers and roles of individuals within them, are not accessible to us. We know that the brain size of *Homo* increased over a period of almost 2 million years, but we do not know with any certainty exactly what such increases mean, and we know virtually nothing about early human social life – nor are we ever likely to. This is what I mean by the lack of balance between theory and evidence. Unrestrained by evidence, the theory is in danger of becoming mere hand-waving – speculation, not refutable conjecture.

Lacking a time machine to take us back 50,000 years or 500,000 years, the best we can do is look for indirect evidence in the form of signs of the postulated evolutionary history that have somehow been stamped upon present-day human cognition. But as we will now see, even such indirect evidence is less straightforward than one might think or wish.

Consider a second line of reasoning: if the social function of intellect hypothesis is even roughly correct, then one of the cognitive skills that is an absolute prerequisite for social transaction, coming before the skill of putting oneself in the shoes of others, is the ability to learn, remember and recognize the features of individuals. Now, we know from studies of patterns of gaze that, when people are interacting socially they look overwhelmingly at one another's faces rather than at elbows, midriffs or shoulders. And when looking at those with whom they are not socially interacting, the gaze is directed at the face even more. In the light of this, it would seem likely that it is predominantly by our faces that we recognize one another. That being so, might not the corollary to the social function of intelligence hypothesis be that facial recognition is a secondary heuristic function that is primed by the primary heuristic? This means that indirect evidence in support of the hypothesis would, for example, be that facial memory should be superior to, say, memory for houses or abstract patterns.

Alas, the evidence is none too clear and, in the end, rather indirect. Some years ago experiments were performed that did show, in a rather singular manner, that there is something special about memory for faces. When pictures of faces are inverted, people have great difficulty in recognizing them, whereas inverting houses, for instance, has relatively little effect on people's ability to recognize them. One possible interpretation of this finding is that we have some form of innate schema for normal, upright faces, but do not have innate schema for houses, and that this makes it difficult for us to recognize faces when they are presented to us upside-down. Supporting this interpretation are the findings of another study, carried out at about the same time, which showed that faces are indeed remembered very much better than the shapes made by ink blots and snowflakes. Yet more support comes from the belief of many developmentalists and ethologists that their observations show human infants to be especially attentive to human faces.

However, all three lines of evidence are open to question and alternative interpretation. The findings on patterns of gaze are unquestioned – when we do look at others, our gaze is overwhelmingly centred on the face. What this means is that there certainly is something special about faces; it is that we spend a great deal of our time looking at them. No other kind of object has such visual familiarity for us. Could it not be, then, that familiarity alone accounts for our being so disconcerted by faces being turned upside-down? There is some support for this. A more recent experiment showed that breeders of pet dogs are less able to recognize animals in pictures that have been inverted, but have relatively little difficulty in recognizing abstract patterns upside-down. Dog breeders, of course, spend much of their time looking at dogs.

But what of the superiority of our memory for faces over our memory for ink blots and snowflakes? The problem with ink blots and snowflakes is that they are highly abstract patterns to which little meaning can be attached. By contrast, other experiments have shown that our ability to remember pictures such as street scenes and landscapes is at least as good as our memory for faces (though inverted landscapes and cityscapes do not present the same difficulties as do inverted faces). Another point to bear in mind about facial recognition is that there is good evidence that this is a cognitive skill much influenced by other features, such as the clothes that an individual wears. So while the common-sense claim for our excellence at facial recognition continues to carry much weight in criminal investigations and trials, for all the reasons outlined above, official reports in the United Kingdom have warned against the widely presumed accuracy of eyewitness testimony based on facial memory and recognition.

Final doubts come from developmental work. The findings of early studies were quite correct: infants do look longer and harder at proper faces than they do either at jumbled faces (where either components like eyes, ears and noses or major lines such as that of the jaw or hair are mixed up) or at abstract shapes. More detailed work, though, subsequently showed that what infants like to look at are not faces as such, but objects that are marked by the features of symmetry, complexity and curvature. Well, OK, you say, but that is simply a crude abstraction of the human face – a face is a complex pattern marked by symmetry and a predominance of curved lines. And it is the human face, with just these characteristics, that is likely to dominate what little visual field infants have. But we now know that what initially really attracts the attention of very young infants is something even simpler: the line that is made either by the eyebrows and bridge of the nose or by the hairline above the forehead. These are what young infants first focus their gaze on. From this simple beginning they start to establish eye contact and to scan the whole face at around eight to nine weeks of age. By fourteen to sixteen weeks the infant has such a good internalized representation of the face that almost any face will elicit a smiling response, and it is only later that the infant shows signs of knowing familiar from unfamiliar faces.

So infants are not born with full-blown a priori knowledge of the human face. Instead they are born with a propensity to pay attention to contrasting horizontal lines; and since it is most often its mother's face which fills the baby's visual field in the first few weeks after birth, it is the way of the infant's world that it is the upper part of the mother's face that first attracts the baby's attention. From this develops a propensity to favour complex, symmetrical and curved shapes, together with a probably instinctive disposition to make eye contact. All this results in the face becoming the focus of early visual attention and the laying down of the foundation for the familiarity with faces that subsequently underlies all later predominant attention, and the ease of visual learning and recognition of faces.

Now, none of this is evidence that the child comes into the world with an innate disposition to learn about faces. None the less, the findings do give some support to the corollary of the social function of intellect hypothesis, that facial recognition has some kind of cognitive pre-eminence. Nature *has* achieved the goal of making the face the predominant basis for individual recognition, but it has done so not with a complete innate schema of the human face but instead with a minimum set of behavioural and attentional predispositions. The regularity of the infant's environment, in which the mother is the dominating visual event, and a brain that responds to accumulating visual experience in a certain way do the rest.

There does seem, then, to be some support for the social function of intellect hypothesis, but it is indeed very indirect. However, it does lead us to a view of the cognitive basis of face learning and recognition that fits well with the adaptations-as-knowledge formulation given in Chapter 4. First, this disposition to attend to and recognize faces, which is a characteristic of all humans irrespective of culture, has a clear goal-directedness that is achieved in a particular way. The internal organismic end of the relationship begins with simple predispositions to attend visually to contrasting horizontal lines and later to make eye contact; this becomes a consistent attentiveness to facial features, which, to a visually competent creature like a human, are an exceedingly rich part of our morphology and hence one of the best bases for individual recognition. The external end of the relationship, of course, is initially those horizontal lines of the mother's face and, later, facial features in general.

Second, the way the adaptation begins as a sensitivity to lines rather than a full-blown innate schema of faces smacks very much of satisficing. It is certainly not the perfect solution, and may occasionally lead to errors of cognition. But the mother's face is a dominating feature of the infant's early visual experience, and the lines and characteristics of curvature, symmetry and complexity to which the infant's visual system is tuned are a pared-down representation of the human face. The system obviously works, and while it may only be a satisficing solution, it is also a rather elegantly economical one.

Finally, what we know of the way this particular cognitive skill grows illustrates well the point emphasized in Chapter 4 as to the way a developmental perspective adds to the adaptation-as-knowledge formulation. It is genes that are partly responsible for the laying down of brain circuits that predispose the infant to pay attention to particular and limited features of its visual world. Subsequent visual experience, and probably further priming by the primary heuristic of the brain, leads to the acquisition of a wealth of detailed information about faces being acquired and stored in a particular part of the brain.

There is one final point to be made about the knowledge-of-faces story. There is a paradox about facial memory: on the one hand we seem to be so good at it, but on the other, when rigorously tested by a legal system that lays great store by eyewitness testimony relating to faces, we seem often to make mistakes. Why is this? One of the reasons so much uncertainty surrounds eyewitness testimony concerning people's faces is that we all have great difficulty describing faces verbally. Try, for instance, to describe your own face or that of someone close to you whom you see every day. This difficulty means that subsequent reconstruction of faces by methods such as artists' impressions or photofit are often woefully inaccurate. We just don't seem to be equipped to describe faces verbally, and indeed we are not. A face is a complex and subtle visual array full of information for a visual system to work with; and in the usual way of things, that is all that is needed. Facial recognition is a specific kind of visual problem solved within the visual sphere, and there is usually no need for us to describe faces verbally.

Now, if the kind of theory developed in Chapter 5 is generally correct, then one would expect intelligence to take the form of skill- or domain-specific *intelligences*, with learning, memory and thinking machinery focused upon specific areas of function. That is, we would expect intelligence to have just the kinds of characteristics that face recognition has: of being 'encapsulated' or separated off from other cognitive skills. In the jargon of modern psychology, the theory predicts 'modularity'; that is, each domain is separated off to some extent from all others, since each domain originates in specific needs of the primary heuristic. Modularity does not claim that this separation of faculties is complete. In many cases it obviously is not. We can indeed say a little about the appearance of someone's face; we just don't seem to be able to convey with language the rich detail that we can detect with the eye and which we use in the everyday task of discriminating between the large number of faces that we all know and can recognize. This lack of total 'translatability' between modules is an important feature of the human capacity to know. And it is explained by the nested relationship of the primary and secondary heuristics.

In this respect, then, human knowledge and its products, knowledge as commonly understood, are not unlike the knowledge gained through the intelligences of other species that have evolved secondary brain heuristics. All intelligence, including that of ourselves, is innately specified and domain-specific. We are no more possessors of a *tabula rasa* than are the birds and the bees.

IS THERE A LOGIC OF SOCIAL EXCHANGE?

All normal humans are rational in the general sense of having the ability to reason. But though rational, there is a large body of evidence that shows we are not very logical, consistent or informed in our reasoning abilities. These deviations away from ideal or 'good' thinking practices take a number of forms.

About thirty years ago, for instance, Peter Wason of University College London showed that people have a strong tendency to reason by asking confirming questions rather than using strategies involving disconfirmation or elimination. He gave people the number sequence 2, 4, 6 and told them that it conformed to a simple relational rule: their task was to discover what that rule was. The procedure was for the subjects to generate three-number sequences and ask the experimenter whether such sequences did or did not conform to his relational rule; when they believed that they knew what that rule was on the basis of his saying whether their number sequences conformed to it or not, then they would announce it to the experimenter. If wrong they would generate further number sequences until they reasoned out the experimenter's rule.

Wason, in fact, used a very general rule – it was 'any series of three numbers of increasing value'. Almost all of his subjects began by generating sequences intended to confirm what they suspected Wason's unstated rule might be – for example, 6, 8, 10 (even numbers increasing by two), or 1, 3, 5 (any numbers increasing by two), or 1, 5, 9 (the middle number is the mean of the outer numbers). Each sequence conformed to Wason's rule, but suggestions as to what that rule was were always made on the basis of confirming instances. Rarely did someone wanting to test, say, the rule of increasing by two begin by generating the disconfirming sequence 1, 3, 4. Had they done so, they would have seen at the outset that, while conforming to Wason's rule, it disconfirmed their hypothesis, which could then immediately be discarded. Subjects generally only came to the use of a disconfirming strategy after repeated attempts at confirmation failed.

Trying to pep up their performance with a considerable financial disincentive for suggesting wrong rules, Wason found that the extra motivation did not rid them of the tendency to use the confirming strategy, but merely made them more cautious in announcing their guess as to what they thought the rule might be; they would generate longer strings of sequences as exemplars of their hypothesis before saying what they thought it was. So here is an instance of most people using a wasteful and none-too-logical strategy; it seems to show a bias in the way that most people think.

Another commonly observed error is considered to be due to what cognitive psychologists call the 'availability heuristic'. The standard example is where English speakers are asked whether there are more words that begin with the letter r (red, rabid, renascent, rotund) or more words with r in the third letter position (strangle, target, error). Only one person in three gets the answer right, which is that words with r in the third letter position are more common. The reason for this, it is thought, is because in trying to solve the problem we try to generate words of the appropriate kind. Because of the way that memory is organized, it is easier to retrieve words beginning with a letter than retrieving words with that letter in some other position. So we sample what is available to our memory and make our (incorrect) judgement accordingly.

Another commonly observed fault in our reasoning is the conjunction error. The laws of probability say that the chance of two independent events occurring together is equal to the product of their separate probabilities. For example, if there is a one in ten chance of getting caught in the rain on the way home (the probability is $\frac{1}{10}$ and a one in fifty chance of having a headache on the way home (here the probability is $\frac{1}{50}$ the chance of getting home wet and with a sore head are one in five hundred $\frac{1}{10} \times \frac{1}{50}$ = $\frac{1}{500}$). Now, few of us are *au fait* with the laws of probability, but everyday experience should confirm that the conjunction of independent events is rarer than one of them occurring on its own. Yet, when posed the problem, most people judge the conjunction as the more likely to occur.

The most general explanation for all such errors of thought is what psychologists call biasing heuristics – of which the availability heuristic mentioned above is one example. In this case the word 'heuristic' is used in a general sense different from that in the rest of this book (and different from its common meaning). For psychologists a heuristic is a simplifying 'rule of thumb', a kind of intuitive statistical inference that we use in place of the complex and detailed logical steps involved in 'correct' or 'good' judgements, which will often require time-consuming and exhaustive searches through memory. So in this specific case, the word heuristic is used to explain a lack of rationality or perfect judgement. We don't retrieve every English word that we know has an *r* in it and then count up those with that letter in the first or third position, and then draw the inference as to which is the more common. Instead, we make a guess, a 'quick and dirty' guess, based on a simple rule. This may lead to errors of fact and logic, but it is quick and quite often works. We usually get by with such solutions, which are a trade-off between accuracy and time and effort. They have H.A. Simon's quality of being satisficing.

The question arises as to whether such rules of thumb are so widespread among people because they are the result of some kind of priming of our reasoning faculties by the human primary heuristic (used in the sense of Chapter 5). An answer may be found in another line of studies on a rather different kind of 'defective' reasoning, but which also begins with certain experiments on human thought originating with work by Peter Wason some decades ago. What Wason did was present each subject with an array of four cards, on one side of which was a letter and on the other a number. Each subject saw one card with a vowel, one with a consonant, one with an even number and one with an odd number. The subjects were told that if a card had a vowel on one side, then it had an even number on the other. The subject then had to say which of the cards should be turned over in order to determine whether what they had been told was in fact true. 'The task proved to be peculiarly difficult,' in Wason's words, even for individuals formally trained in logic.

The 'correct' − the logical − way of solving the problem is to check both the card displaying the vowel (to confirm the truth of the statement) *and* the card showing the odd number (to see whether the truth of the statement is disconfirmed). Most subjects did indeed check the card showing the vowel, and rightly ignored the one with the consonant (everyone realized that the card showing a consonant had no relevance at all to the truth of the statement that had been made, which concerned only cards with vowels). However, almost everyone made the error of not checking the card with the odd number, choosing instead (and in error) to check the one with the even number. This latter response is an error because the initial statement did not say what a card with an even number must have on its reverse side − only that if a card shows a vowel then it must be paired with an even number. There may well have been cards with an even number on one side and anything at all on the other, but such cards would not have affected the truth of the original statement.

The Wason task provided four possible cues: a vowel (P), not a vowel (not-P), an even number (Q) and not an even number (not-Q). The complete logically falsifying response is 'P and not-Q' (that is, check the vowel and the odd number in the form that

Wason presented the problem), and only about 10 per cent solved the problem correctly. However, the propositional content of P and Q can be varied, and that is when it becomes interesting.

A well-known example comes from a study carried out in the United States about ten years ago. Subjects were told that they must imagine themselves as bouncers in a Boston bar and that their jobs would be in serious jeopardy if they did not enforce Massachusetts law, which states that only persons over twenty years of age can drink beer. They were then presented with cards bearing on one side one of two kinds of drink being consumed, and on the other one of two ranges of age. They were told that the cards represented possible situations that they, as bouncers, would encounter, and they were asked which situation must be investigated further (that is, which cards must be turned over) to make sure that the law was not being broken and hence that their jobs were safe. The cards were drinking beer (P), drinking coke (not-P), more than twenty years old (Q) and less than twenty years old (not-Q). In contrast with the 10 per cent level of correct solutions found when the contents of the cards were uninteresting neutral letters or numbers, when the propositional content took the form of the age–drinking problem, 75 per cent of people responded correctly with both P (check the age of the beer drinkers) and not-Q (check that the under-twenties are not drinking beer) – realizing that the age of coke drinkers (not-P) and the drink of those more than twenty years old (Q) were irrelevant to the legal situation. So despite an identical logical form, the problem is usually properly solved if put in the meaningful form of how bouncers should behave, but is not solved properly when presented in an abstract manner.

This particular story has now been taken one step further by L. Cosmides, a psychologist in the United States, who has provided an even more revealing and interesting variation on the Wason task. In order to understand what Cosmides did, it is important to understand her theoretical position, which is not too dissimilar from the general position of this book. Human beings, Cosmides asserts, are not general problem-solvers and have not evolved an ability specifically to think about and solve arbitrary logical problems – such as that of the original Wason task. As thinkers, humans are products of a particular evolutionary history:

Our species spent over 99% of its evolutionary history as Pleistocene hunter-gatherers: the genus *Homo* emerged about two million years ago, and agriculture first appeared less than 10,000 years ago. Ten thousand years is not long enough for much evolutionary change to have occurred, given the long human generation time; thus our cognitive mechanisms should be adapted to the hunter-gatherer mode of life, and not to the twentieth-century industrialized world.

We think adaptively rather than logically, and what adaptive means for Cosmides is what would have been adaptive a very long time ago. This 'ancient' way of thinking is thought to have been moulded by the requirements of social exchange; and the notion of social exchange is a variant on the social function of intellect hypothesis.

Social exchange means that social interactions are regulated by the rule that individuals must always pay a cost, or meet some form of requirement, in order to receive a benefit. For example (though not one used by Cosmides), one may choose a wife from a community (the benefit) only after delivering to that community the fresh corpse of a sabre-toothed tiger (the cost). A crucial part of this 'logic of social exchange' is that we are especially sensitive to cheating, that is, to detecting individuals who fail to reciprocate on social contracts. Specifically, we should be predisposed to be on the lookout both for people who are enjoying a benefit (find out whether they have paid the cost) and for those whom we know have not paid the cost (do they enjoy the benefit?). In the language of this book, we humans should be predisposed by our primary heuristic to have secondary heuristics that are especially sensitive to the cost-benefit propositions of social exchange.

Well, are we? asked Cosmides. Consider social exchange in the rather deathly logical language of Ps and Qs of the Wason task. The four cards are as follows: 'has benefit' (P), 'does not have benefit' (not-P), 'has paid the cost' (Q) and 'has not paid the cost' (not-Q). How would we, as law-enforcing elders of some palaeolithic group of hunter-gatherers, check that no one was violating the 'wife for sabre-toothed tiger' agreement? We would do that by making certain that those with wives have indeed delivered on the tiger deal (P), and that those who have not produced tigers do not have wives (not-Q). Once someone has delivered a tiger (Q) his marital status cannot be one of cheating; and if someone does not have a wife (not-P) then he cannot violate that particular social contract. So, in this case, the logical falsifiers P and not-Q coincide with the correct strategy of checking for cheaters. Of course, the social lives of early people were richer than this, presumably extending to the likes of special eating privileges gained through protecting the group from hostile neighbours and sexual favours granted through sharing foraged fruit and honey, to mention but two. However rich the network of agreements and contracts, though, the logic of social exchange remains the same.

Now, what Cosmides did was present the Wason task to people, but dressed up in the propositional language of something like the wives-for-tigers scenario. In fact, her experimental examples drew heavily on imaginary situations involving foraging and food sharing, because social exchange in early human history most likely revolved in large part around interactions concerning the gaining and sharing of food resources. She wanted to test one explanation for the apparent excellence of people's thinking when tested on the age–drinking version of the Wason task, which contrasts so startlingly with the poor performance of people when the task is presented in an abstract form. We have, in fact, already encountered this explanation some pages back when we had cause to refer to the 'availability heuristic' as one of the rules of thumb that biases our thinking. In this instance the argument is that drinking and age are frequently linked in our culture. As a result of this much-used association, we are better equipped to think properly through the Wason task when its propositional content refers to this, or indeed any other already associated concepts.

Cosmides is hostile to explanations like the 'availability heuristic' because such a mechanism is not domain-specific. As a general process or mechanism it smacks of the blank slate. Thus parts of her experiments are aimed at testing such an explanation by providing a propositional content for the Wason task that is at once couched within the language of social exchange but also unfamiliar – few of us have any experience either of tigers or of their being exchanged for the benefits of marriage. In this way Cosmides provided a first test for her idea that a problem which is domain-specific in the form of the logic of social exchange will not be influenced by a lack of familiarity. Her data supported her hypothesis. Most people solve the problem in a way that is both logically complete (P and not-Q) and correct in terms of the logic of social exchange (check those with wives and those who have not delivered tigers). Since the material is unfamiliar, she argued that her findings cannot be interpreted in terms of the availability heuristic.

Cosmides had a further clever test of her hypothesis of domain-specific thinking. As just mentioned, when the propositional content of social exchange is presented in the form, 'If you want a wife (P) then you must first deliver a tiger (Q)', the rules for spotting cheaters and the logically correct falsifiers are both P and not-Q. So what Cosmides also did was present to people what she called the switched social contract, which took the form, 'Those who deliver a tiger (P) will have a wife (Q).' Here the logical falsifiers remain the same (P and not-Q), but the requirements of looking for cheaters require not-P (if you have not delivered a tiger then you should not have a wife) and Q (if you have a wife then you must have delivered a tiger). The logical falsifiers P (having delivered a tiger you cannot possibly be a cheat) and not-Q (if you haven't got a wife then you cannot be cheating), are irrelevant to the search for cheats. So in this case the logic of social exchange diverges from that of formal logic. And what her subjects provided were responses that supported the predictions of the logic of social exchange and not those of formal logic.

I have presented here a mere thumb-nail sketch of an extensive study, which has not been without its critics. Innovative science which leads to results contrary to those expected by more orthodox predictions is always criticized – this is, in fact, an important characteristic of science. As will be argued towards the end of this chapter, science is a selectional process, like all other forms of evolution, and the caution that comes from an entrenched orthodoxy is an extremely effective selection filter. New ideas that survive in the face of establishment opposition become incorporated into the corpus of established science and serve in turn as a selection filter for later ideas. Cosmides's work has been replicated, and must be replicated and extended in cultures very different from ours, because one of the demands of the notion of secondary heuristic priming by the primary heuristic is that it is to be found in all humans whatever the current cultural demands on their thinking might be. My prediction, not surprisingly, is that such replications will be made and that something like the logic of social exchange will be incorporated into mainstream cognitive science.

Domain-specific, innately constrained forms of reasoning (corresponding roughly to the processes and mechanisms by which we gain knowledge) are no more unexpected than domain-specific, innately constrained memory (corresponding to knowledge itself). We may like to think that we can think of anything and in any way we choose. And perhaps we can, but doing so takes a great deal of hard work. For most people logic and mathematics are not 'easy'. Unconstrained, general, context-independent and domain-unspecific thought does not come naturally.

LANGUAGE

A question that we all consider at some time is, what is it that makes us different from other animals? The use of tools, the realization that others have minds, the quality of mercy and compassion in our treatment of fellow human beings, and culture have all been argued for as what makes us human; and all have been observed or attributed in some usually partial and uncertain way to other species. We do, though, seem to have one ability which is truly unique to ourselves: language.

Certain other species of animal can transmit complex information with considerable precision, the most famous example of which is the dance of bees on returning to the hive, which transmits to other bees the position and richness of food sources or a promising new hive site. And it is now known that some other species have a limited ability to use specific reference in their communication; that is, they have signals that correspond to specific objects in their environments. For example, a common African monkey, the vervet, has different signals for flying predators, ground predators and snakes; and other vervets hearing these signals react in ways appropriate to the danger being signalled.

Impressive as such feats are, they are not language. Language is the ability to use a relatively limited – albeit quite large – set of symbols to generate a virtually infinite number of meaningful combinations to form utterances, each of which has meaning. I can, for example, say, 'I am going to France', and 'It is to France that I am going', and 'I am on my way to France', and 'France here I come', and many others besides, all of which have the same meaning. Each of us has a vocabulary of around 30,000 to 40,000 words, of which the most commonly used number just a few thousand. Yet we are able to generate some 10^{30} (that is a 1 with thirty zeroes behind it) different meaningful sentences of about twenty words or less in length. Now, 10^{30} is an extremely large number – so large, in fact, that it dwarfs the 10^{10} seconds which is the approximate length of a normal human life span of threescore years and ten. (Ten to the power of 30 is not three times larger than 10 to the power of 10; it is twenty orders of magnitude greater, that is, a number with twenty more noughts in it.) If we never drew breath but just talked all our lives, we would utter but a tiny fraction of all the sentences we are capable of producing. Of course, many of these would be novel – they will never have been produced by us, or others, before. Indeed, if we look at the previous two or three sentences on this page, or at the sentences following, it is more than likely that I have never produced them in that exact form before in my life. It is *this* characteristic of human language, its extraordinary creativity, that makes our linguistic ability different from anything that any other species of animal has.

There has been a small storm among psychologists and linguisticians going back about thirty years as to whether apes, especially chimpanzees, can be trained into language use. This work has its beginnings in older, and very curious, experiments in which infant chimpanzees were brought into human households and raised, effectively, as if they were human children, as one of the family. They spent the same amount of time with their human 'parents' as ordinary children would; they were fondled, played with and *spoken* to as one would fondle, play with and speak to human children. Under these circumstances the ability of such animals to develop any sort of spoken human language was virtually *nil*. They just did not learn to speak. Now, this may merely have been because chimpanzees lack the kind of vocal cord apparatus, and the neural control of such apparatus, that we humans have. Perhaps these chimpanzees indeed learned to use language, but could do so only in their heads, lacking the appropriate peripheral anatomy by which their language could be expressed vocally.

In other words, these early experiments were bad experiments, and the reason for their failure, perhaps, is a trivial one. This possibility led to the question being asked as to whether exploiting some peripheral apparatus which chimpanzees do have, their hands, might not lead to more interesting and revealing results. Chimpanzees are very good at using their hands. Might not sign language, of the kind used by deaf people in the United States in which signs stand for whole words, be the symbolic medium by which chimpanzees might come to learn and use language, just as deaf people do? A quarter of a century of experiments lead us to the answer 'no'. Such apes certainly do acquire something in the region of 100 to 500 hand signs (symbols), most of which are for objects in the world, a minority being action words (verbs) and a limited number of action modifiers (for example, words like 'more'). But no good evidence has ever been obtained that these animals have ever used these symbols in the structured, organized way (that is, syntactically) that characterizes human language; and no evidence at all exists for the creativity of symbol use, that is the *sine qua non* for accepting that it is language that one is observing in these animals. Whatever one might think of these experiments, they certainly do not demonstrate the existence of linguistic ability in these creatures.

Given that language is unique to our species, that must mean that some part of that portion of our genetic makeup that is unique to us as a species is a part-determiner of the human ability to function in the realm of language. In other words, we must be genetically predisposed to learn, think in and communicate by language. This uniqueness of language as a human form of knowing about the world, as well as its apparent multiple roles in human psychology, means that an understanding of language is an essential part of a science of human knowledge. We come to know a language in the sense that language is a part of our environment, it is something 'out there', as important and prominent as any other set of features of the environment. Also, to some extent – and some people think it is to a very considerable extent – we think using language and hence manipulate our knowledge of the world through language. We also communicate our knowledge to one another primarily through language.

This nexus of roles for language means that the relationship between internal organization and external features of the world that characterizes all forms of knowledge takes, in the case of language, a dynamic, double form: a word, or sentence, as external event is matched to internal representations of the meaning of the utterance, and meaning itself may take more than one form; and we also have words, or combinations of words, in our heads which match a very large number of features in the world. It is, of course, this symbolic quality of language that makes it so central to our psychology: a dog and the word 'dog' are both 'things' in the world; and our internal representations also have this dual quality of some kind of image, primarily visual perhaps, of the dog, the thing itself, and the brain state that corresponds to the word 'dog'.

On the other hand, we should be careful not to overstate the power of language in our knowledge of the world. We have already seen how limited we are in dealing with faces using language; and in the next section we will see that there is a general poverty of language when it comes to dealing with emotional states. And yet, language so dominates our lives, especially our interior mental lives, that it is difficult not to think of it as anything other than the central plank of human understanding and knowledge.

So, returning to the main thread of the argument, from the simple fact of the uniqueness of language to our species, the case can be made, though not an absolutely watertight case, that we do not come to know language itself, or to know the world through language, in the same way and via the same mechanisms that members of other species come to know, say, the spatial relationships of their worlds or how they come to associate temporally related events. Knowledge of, and by way of, language cannot be accounted for by some generalist, non-species-specific mechanism. But can the argument be made that knowledge of a language is also not attributable to some species-specific yet generalist ability of the human mind? The answer is yes. There is strong evidence for language being understood as a genetically determined and domain-specific organ of mind that goes beyond its merely being unique to humans and the difficulties that we have in describing faces using language or giving adequate verbal expression to our emotions. In other words, once again the evidence seems to stack up against a *tabula rasa* approach to human knowledge.

Consider, for example, some further remarkable figures about language learning and language use. At one year of age all normal human infants, irrespective of which of the 5,000-plus languages of the world they are being raised to speak, have a vocabulary of about ten words. Four years later the average vocabulary of five-year-olds is somewhere between 5,000 and 10,000 words. Such an increase requires the learning of about five new words each and every day throughout that period. This is an extraordinary achievement, because it occurs in such an unguided, untutored, unintended way. Adults and older children do not drill younger children into learning words and their meanings. They learn by a process that seems, by its ease of occurrence, analogous to diffusion – words and meanings seem to seep from the linguistically rich environment surrounding the child into the brain of the child. It is, as has been remarked by others before, as if children cannot stop themselves learning a language. It takes very little to stop children learning to read, write or ride a bicycle. But language seems to have the magical quality of largely growing from within the child – not completely, though, for those rare instances of children raised in environments totally devoid of language show us that such children do not have language. But given some minimal linguistic component in a child's world, in a relatively short time that child will come to a mastery of language.

Mastery of language does not just mean having the phonetical know-how for pronouncing words properly and a vocabulary to go with it. It also means speaking with the correct syntax – that is, the rules that govern word order, and the wider grammatical rules governing forms such as word inflections (these are changes in word endings in accordance with tense or to indicate plurals). For example, by five years of age every English-speaking child knows how to say, 'Give the dog a bone.' No normal child would ever produce sentences like 'The bone dog a give' or 'Bone a the dog give'. And all children of that age know how to transform a declarative, such as 'John is at home', into the interrogative form 'Is John at home?' Later transformational tricks will include such linguistic wizardry as effortlessly turning active constructions ('John gave the dog a bone') into passive ones ('The dog was given a bone by John').

What is extraordinary in all such examples is the contrast between the casual immersion into a language environment which the children experience and the highly specific and intricate language structures that emerge. The parents and caretakers of children are usually quite permissive in what they allow young children to say. They may occasionally correct gross errors, such as an incorrectly generalized past-tense rule like 'John gived the dog a bone', but they often let these pass, and it is not explicit lessons in grammar that result in the child becoming linguistically so capable. Specific, disciplined learning simply cannot explain the way in which language develops in the child. Once again, as in the case of birdsong discussed briefly in the previous chapter, language learning in the human child is well described by the seeming paradox of the child knowing what it has to learn. All attempts at accounting for linguistic development in the child by using 'generalist' learning processes and mechanisms such as associative conditioning and trial-and-error learning are now wholly discredited. One can no more explain the exquisite structure of a five-year-old's language just by the environment in which it has been raised any more than one can explain the structure of that child's hand just by the meals that it has consumed or the objects that it has grasped. In both cases, language and hand, the structure comes in large part from genetical information that is nurtured and finds expression in a supportive, but not directly instructive, environment.

This contrast between an impoverished and unstructured environment, on the one hand, and the specific and intricate structure that develops *in all children*, regardless of differences in their upbringing, on the other, is what Noam Chomsky, the American linguist, called 'the argument from the poverty of the stimulus'. The argument is taken as very strong support for the assertion that language is caused by 'innate factors [that] permit the organism to transcend experience, reaching a high level of complexity that does not reflect the limited and degenerate environment', as Chomsky himself put it.

There is yet further evidence for the innate determination of linguistic learning in the stages of language development that occur in all children, again irrespective of the language that they come to speak in time and irrespective of the culture in which they are raised. All children begin to speak with one-word utterances at about ten or twelve months of age. These are made up predominantly of object words (such as 'ball' or 'daddy'), with lesser numbers of action words (for example, 'throw' or 'hug'). All children then produce two-word utterances at about twenty to twenty-four months. These two-word utterances are structured (in English, for instance, the infant might say 'throw ball', almost never 'ball throw'), without functors – that is, without prepositions (like 'on' or 'in'), conjunctions ('and', 'but', 'or'), articles ('a', 'the'), auxiliaries ('have', 'has') and copular verbs ('am', 'is', 'are', 'was') – and without inflections; functors always appear later. This order is almost never violated, and, again it is worth stressing, the universality of the pattern of development occurs irrespective of the child's specific experiences.

There are other universals of human language that are present, irrespective of the language spoken or the learning experience. One such is the oddity known as 'motherese'. 'Motherese' is a misnomer, being a form of speech adopted by all adults when addressing infants: the baby seems to 'release' a special form of slow, high-pitched talking with exaggerated intonation ('koochee-koochee-kooo' or its equivalent in Japanese, Swahili or any other language), a form of speech that adults never normally use when addressing other adults or even quite young children. There is also evidence that infants pay more attention to 'motherese' than to ordinary speech. Less of an oddity, and just as fascinating, are studies carried out on deaf children raised using hand signs rather than spoken language. Such children pass through exactly the same developmental stages as speaking children, and the language that emerges has the same characteristics of grammatical structuring. So language is not confined to vocal speech and the auditory channel. It is a 'knowing' part of the human mind that normally attaches itself to tongue and ear, but functions almost as well when operating through the hands.

Finally, linguists have for decades been trying to uncover the universal transformational rules that perhaps govern the structures of all languages, and through which all languages can eventually be described. One rule that does seem to be universal is used in the transformation of a sentence from the declarative to the interrogative form. For example, 'The tiger is alive' becomes 'Is the tiger alive?' The rule is that without regard to any specific language it *always* involves a syntactic restructuring of transposition with the verb following the noun phrase. So 'The tiger that is stripy is alive' becomes 'Is the tiger that is stripy alive?', and never 'Is the tiger that stripy is alive?' or 'Is stripy is alive the tiger that?'

Since the late 1950s, the work of Noam Chomsky and others has completely changed the way that psychologists understand human language. It has come to epitomize the general approach that human knowledge is innately determined and domain-specific. This Chomskian understanding of language fits well with the picture of the evolution of intelligence developed in Chapter 5. *Homo sapiens* has a spectacularly evolved capacity for learning and remembering an extraordinary amount of linguistic material, and of creatively generating a prodigious quantity of linguistic output. This is a tool both for dealing with rapid change by thought and communication and for generating change in the form of a rapid flux of ideas and conceptions. What I cannot yet understand, and neither can anyone else, is exactly what the functional origins of language are. If, as argued in the previous chapter, instincts are the mother of intelligence, we are not yet able to understand the nature of that originating instinct. Perhaps it arose as part of the social function of intellect à la Humphrey and Cosmides, with language supplying symbolic form to more concrete types of social exchange and social contract, the symbolism giving rise to possibilities of greater complexity and subtlety of interaction. Or perhaps it arose in response to, or hand in hand with, the requirements of the evolution of human culture, a tertiary heuristic that can operate at even higher-frequency levels of change than the secondary heuristic of individual intelligence, and hence which requires a medium of information-transmission that can match those high frequencies of change.

It is with culture, and the obvious place of language in culture, that I want to complete this briefest of surveys of human knowledge. But before then I want to comment on an area of human experience and knowledge in which language seems to have a much reduced role. This is the knowledge and experience of emotion.

EMOTIONAL KNOWLEDGE

If there were no adaptations, there would be no life. Since adaptations are knowledge, this means that without knowledge there would be no life. Knowledge is a fundamental characteristic of life. Life depends on knowledge. That has been one of the central arguments of this book. But life is not *about* knowledge. Most evolutionary scientists now believe that life is about genes and the perpetuation of genes in time. Knowledge serves genes, whether that knowledge takes the form of adaptations supplied by the primary heuristic, or knowledge as more commonly understood, which is a product of intelligence, given to us by the operation of the secondary heuristic.

In the terminology introduced in Chapter 3, the gene is the unit of selection – the replicator – of that most fundamental and irreducible level of the hierarchy of all evolutionary processes. There are other units of selection, other replicators, at less fundamental levels, but because of the nested relationship of the hierarchy, the gene is the central evolutionary mechanism around which these satellite evolutionary processes and mechanisms revolve. In the language of Chapter 5, the secondary heuristic does the primary heuristic's bidding; the secondary heuristic provides the knowledge that the primary heuristic itself is too slow to establish. The priming of the secondary heuristic has been considered until now to be a matter of the primary heuristic 'pointing' to what specific features of the environment must be attended to, learned and reasoned about. It does this because of the nesting of the secondary heuristic under the primary heuristic. I want now to consider another possibility. This is that the primary heuristic makes itself felt by an additional and rather more diffuse device – that is, by giving rise to feelings, emotions, that act to guide further the operations of the secondary heuristic. Put in ordinary language, emotions are there in order to tell us what to think about; our hearts not only try to rule our heads, but should perhaps be allowed to do so.

Consider again the way we first considered the matter of emotion in Chapter 1. Magna Carta was signed in 1215: I know that as a simple 'fact' without any emotional content at all. The year dates that mark the birth of my children, on the other hand, I know both by their factual content and by the emotional feelings that I observe in myself when I think of those years. If knowledge is to be thought of as the relationship between internal organization and external features of the world, then a not insignificant amount of knowledge as commonly understood is marked by certain feelings, emotional feelings, as part of the internal end of that relationship. For convenience, we can call such knowledge emotional knowledge, which is to be distinguished from other forms of knowledge precisely because certain bodily states accompany emotional knowledge.

The suggestion, then, is that emotions are mental and physiological states (resulting in sensations like a churning stomach or a bursting feeling in the chest) that signal the possible or actual presence of biologically significant events in the world, their significance having been gained by the primary heuristic over long periods of time. The signalling might be direct and actual – for example, when feeling pain caused by tissue damage. Or it might be indirect, possible and routed via the secondary heuristic, as in the case of pleasurable events or sensations being linked to someone with a certain face or smell. Either way, events in the world are signalled as being good or bad, those which we should attempt to attain or avoid, those associated with life and those associated with death. The last mentioned, that most fundamental of dichotomies, is the most direct message that we can get from our genes as to what to do, what to avoid or what is worth investing some secondary heuristic capacity in; but it doesn't have to be as dramatic and stark as life and death. Pleasure and pain are signals too, if less intense and less dramatic, but ultimately deriving from the approach–avoidance or life–death dichotomy.

Contemporary psychology actually postulates the existence of at least six basic emotional states (happiness, surprise, sadness, anger, disgust and fear). All must ultimately derive from the fundamental dichotomy of what perpetuates genes and what does not. Also, given the approximate correctness of six basic emotions, from combinations of these we can arrive at a very large number of emotional chords or colorations for the events in our world. But, whatever the emotion being signalled, one of its functions is to tell us what to attend to, what to learn about. This is a very conventional view of emotions dressed up in modern evolutionary terms. Emotions are postcards from our genes telling us, in a direct and non-symbolic manner, about life and death. The only novelty in the approach adopted here is that it is couched in the terms of the primary and secondary heuristics. The primary heuristic is complemented by the additional information of emotion in the way it informs the secondary heuristic about which predictably unpredictable features of the world it should concern itself with.

Emotional experience is rich and subtle. However, except in the hands of people with a special talent, more of which in a moment, it does not lend itself to easy verbal expression – except in the crudest terms. Once again we have an example of the encapsulation, the domain-specificity, of cognitive function. The way in which most people express and communicate emotional states is known as non-verbal communication. Non-verbal communication is something that we do without any of the necessary intention to communicate that underlies verbal communication. Telling someone that you love them, the verbal communication, is quite different from showing someone that you love them, the non-verbal communication. In the former case it is a simple utterance. In the latter case one exhibits certain uncontrollable bodily states such as breathlessness, trembling extremities and other manifestations of autonomic nervous system function. (The peripheral nervous system is divided into a somatic part, which controls skeletal muscles and gains information from skin and muscle, and an autonomic component, which controls, and receives sensations from, viscera such as heart, lungs and gut as well as genital organs. The central nervous system of the spinal cord and brain is rather more closely integrated in function, but still roughly separable into these component parts.) Other important components of non-verbal communication consist of posture, hand gesturing and facial expression, including size of pupils. Only actors and, as the anthropologist Gregory Bateson liked to point out, confidence tricksters (con men), are adept at controlling these non-verbal signals of emotional state. For the rest of us, non-verbal communication is largely beyond voluntary control. We cannot deceive with this signalling system as we can with spoken language. And that is why lovers are apt, quite rightly, to demand that they are shown that they are loved and not just told it.

Emotion may thus serve two cognitive functions, not just one. It informs our intelligence what to learn and think about. And it also serves as a reliable signalling system of how we are going to behave towards others, or how they are going to treat us. But however rich both the experience of emotion and the signalling of it through non-verbal communication might be, it is a very different form of experience and communication than what comes to us through spoken language. With a few obvious exceptions – so exceptional that we give them strange technical names like onomatopoeia (words that sound like the thing being signified, such as 'bang' or 'murmur') and synecdoche (metaphorical use of part of the signified in place of the whole, for example 'head' meaning a whole cow, as in 'I am moving a thousand head through the pass tomorrow') – spoken language is highly abstract in the sense that the symbols seldom correspond to the signified, and hence also there is very little part-for-whole symbolism.

Emotions, by contrast, are expressed and communicated in large part through iconic, part-for-whole symbolism. A clenched fist is an involuntary response to the emotion of anger, and is communicated as a signal, warning of a possible blow to come. Here the clenched fist is taken out of the whole sequence of physical attack, and comes to stand for and signal the emotion of anger that accompanies an attack. It is, of course, possible to tell someone that you are angry with them and to have them believe you. It is far more convincing and effective, though, to act as if you are indeed angry, so angry that you teeter on the edge of physical violence. The clenched fist, the pallor and the trembling frozen stance are a qualitatively different kind of message than the statement that describes it. The same differences hold for the dilated pupils and sighing demeanour that accompany, and signal, love and happiness; or the contorted, rejecting features of disgust. The actual or even blown kiss carries an impact that can never, or almost never, be conveyed in words.

This separation of emotions from spoken language is similar in kind to that between facial recognition and our poor ability to describe faces in spoken form. Just as facial recognition has a specific role to play in ordinary human life which facial description does not, so emotions have evolved to do a different job from spoken language, and they do it in a different way. It would not do to overstate their separateness. One might, when suddenly confronted by a man with a gun, shout to one's companions, 'I'm afraid!', which will lead to rather pointless consternation and enquiry; one is much more likely, and effectively, to shout, 'Look out! There is a man with a gun!', with the tone of voice carrying the message of fear and the abstract word 'gun' doing what spoken language does with such wonderful effect, namely serving to signify some specific object in the world. In this way we mix verbal and non-verbal communication all the time in our daily lives. But the emotional content is almost always present. Flattened affect is a sign of pathology. Normal human life is lived within a sea of experienced and expressed emotions; emotional knowledge may be different from, but is every bit as important as, other forms of knowledge.

In Chapter 3 it was suggested that the way that scientists divide the world into significant and insignificant parts is seldom congruent with everyday experience of which bits of the world are important and which are not, and that this is one reason for the seemingly 'strange' feeling that non-scientists have about science. There is another. The more abstract and precise the language of science, the happier is the scientist. All science begins with description in simple, everyday language; reasoning in the form of categorization then appears; and this is followed by ever-more-remote language forms, jargon, that express knowledge that is specific to that science and are increasingly mixed with formal logic and mathematics. This remote abstraction, so very different from the iconic, overlapping mixtures of emotion that give ordinary life its rich flavouring of emotional substance, is another reason why science is sensed as being so odd, so dull, so 'unnatural', by the non-scientist.

By the same reasoning, ordinary spoken language is a rather bloodless account of life, except in the hands of that special group of people that we mentioned earlier. Since the beginning of time, story-tellers have been honoured precisely because they have the remarkable gift of rendering in words those postcards from the genes. Being able to say in words what the rest of us can only experience and communicate with our bodies is an extraordinary talent. It is what separates literature from science and makes literature art. Science aims to establish knowledge of the world in which emotional knowledge has no part. Art seeks to understand the world through emotional knowledge. These are separate enterprises and should be kept so. Scientific knowledge can inform art and be its toy, a source of ideas. Something like this is beginning to happen as scientific notions begin to appear in contemporary fiction. But it is a mistake to think that science can be used interchangeably with art. The 'two cultures' of science and art have their origins in the domain-specificity of all human knowledge. It is not that they aren't 'best' kept apart, but they are apart: they are different forms of knowledge.

CULTURAL KNOWLEDGE

Towards the end of his life Piaget wrote a book the original title of which, roughly translated, was 'The Behavioural Motor of Evolution'. In it he conferred very considerable significance on the role of behaviour in the evolutionary processes that lead to adaptation and speciation, that is, the primary heuristic of this book. Alas, he did it within a Lamarckian framework. We know that Lamarckianism is wrong and therefore we can safely reject Piaget's views about biological evolution, however much respect one may accord his other writings. Yet what, in fact, Piaget had attempted in some detail was but an extension of the ideas of many nineteenth-century evolutionists as well as such notable twentieth-century biologists as Konrad Lorenz and Ernst Mayr. They had also attempted, if much more casually and within the confines of an orthodox neo-Darwinian approach, to make the case for including behaviour, along with genetics, as one of the causes of evolution.

The attractiveness of the idea that the behaviour of individual animals may have significant causal power in the evolution of the species to which those individuals belong is not too surprising. There seems to be something different, something much more dynamic, about the behaviour of a blackfooted polecat when compared to those black feet. However, I argued in the previous chapter that such views are incorrect. They are a kind of delusion that behaviour automatically adds dynamism to the evolutionary process. As far as the evolution of species is concerned, behaviour is no different from any other trait of the phenotype. Behaviour only becomes causally significant in the process of evolution if the behaviour is driven by intelligence, itself an evolutionary process involving unpredictable variant generation. Then and only then is it impossible to understand biological evolution within the reductionist terms of genetics and development.

Since a fair proportion of human behaviour is driven by intelligence, then it follows that in the case of our own species, what happens in our heads has been of real importance − of causal importance − in the evolution of *Homo sapiens*. Whether our brains have been of equal, lesser or greater importance than our gonads in human evolution is a pointless argument that cannot be resolved. But that both brains and gonads have carried causal significance in that evolution seems to be indisputable. One aspect of human intelligence in particular has most often been invoked as having had evolutionary significance, and that is the ability that allows us to enter into the collective form of knowledge that we call culture, which in Chapter 1 was referred to as shared knowledge.

Shared knowledge takes many forms. One might be the understanding gained by our colour-blind person of Chapter 1 from elementary physics texts explaining about the scattering of sunlight by the atmosphere which results in his or her knowledge that cloudless skies are blue. Another might be knowledge gained from a television advertisement that a new soap powder is available, or that skirts this year are above or below the knee, or that some new horror is being perpetrated against people in some distant part of the world that most of us will never visit but all of us know about. In all such cases, and indeed in all of education, the common element is that knowledge gained by one person's or several persons' secondary heuristic of intelligence is transmitted by a learning process to another person or other persons. Culture is learning about what others have learned, created or invented.

Whatever the contents of culture, the bits and pieces that make it work are, initially, innovation or innovative learning of some kind, and then individual learning and memory, and the means of communicating that learned knowledge. Cultural communication is the means of spread or transmission of knowledge and usually occurs via language. Non-verbal, individual observation of the learned behaviour of others certainly would, and does, result in cultural transmission, but it is a much more limited case. This slow spread by way of one individual watching another individual is more typical of animal protocultures, as occurs in the case of novel food-finding or food-accessing methods such as breaking into nuts or opening shells; and it is also common in the learning of motor skills in early human childhood, like those of dressing or manipulating implements of various kinds. But in the case of a man going to the moon and imparting to the rest of us what it is like to walk upon it, or of one who visits a mountain top and descends to spread the word of God, or in the dissemination to other scientists of the structure of a gene or the solution to a mathematical problem – in all such cases language is the means of transmission.

A feature of cultural knowledge that often distracts people from a real understanding of culture, of how it works and what it is, is that it can be stored exosomatically. This means that it can be stored outside of our heads, in books, on microfilms, on magnetic tapes or on computer disks. With some possible exceptions to be mentioned later, this is largely a modern irrelevance, because culture is probably almost as old as our species and certainly as old as language. Culture, in short, has been a potent force in human life since long before exosomatic storage of any kind was invented.

What is not an irrelevance is the content of culture, because, for reasons that we do not yet understand, content determines the stability of cultures and subcultures. By stability I mean the rates at which the items of knowledge that are transmitted between individuals are changed, revised, added to or abandoned. The anthropologist W. Goodenough defined culture in functional terms as 'whatever it is one has to know or believe in order to operate in a manner acceptable to that society's members'. However, the 'shelf life' of such knowledge and beliefs that allow competent performance within a community is extraordinarily varied. Consider, for instance, the obvious contrast that is presented by dress fashion on the one hand and a specific religious belief on the other. Both may fulfil Goodenough's definition in terms of what must be known in order to be accepted by others. But whereas religious belief has spanned millennia and thereby exhibits astonishing stability, sometimes in the face of great pressures to alter or abandon such beliefs, fashion in dress in Western societies is notoriously and predictably fickle. It is likely that such differences are due to differences in culture-level selectors, an issue that will be returned to shortly. First, though, we need to understand how such a concept as cultural selection can be invoked.

In Chapter 3, we developed the fantasy of Darwin as, among other things, a scientist who discovers the principles and processes of evolution by selection through the study of a particular kind of culture, a community of scientists. The point of the fantasy, of course, was that it was not quite as fantastic a tale as it might at first seem. Indeed, in recent years biologists with an interest in culture have come to assume that cultural change is the result of the same evolutionary processes as those responsible for evolution as more conventionally understood. The mechanisms are obviously different – that is, the things that we must put our fingers on if we were identifying the causes – but the processes of variation, of the selection of a small subset of variants, and their propagation or transmission are the same. In *The Selfish Gene*, Richard Dawkins coined the term 'meme' as the cultural equivalent of the gene. The meme is the cultural replicator. Using the terminology of the meme, current thinking about cultural change can be summarized as follows.

Memes are roughly equivalent to ideas or representations, that is, the internal end of the knowledge relationship. Memes take variant forms. The variation may occur *within* individuals (my brain state that represents an external environmental feature, a complex one in this case, of 'market forces', for instance, may vary a little from day to day depending upon intervening experiences and what else I have been thinking of, or it may undergo more drastic change in the light of experience, which is likely to be a more important form of variation); and they may vary *between* individuals (in one sense it is unlikely that any two people have identical thoughts about market forces – indeed, it is likely that we all have different internal states for all possible external features of the world, though doubtless these could be roughly grouped in certain ways into similar classes; in a more important sense there may be substantial individual differences of belief). Selection acts upon these variant memes directly; for example, I discuss with a friend the notion of market forces and am convinced that my ideas in this regard are wrong and should be altered to match more closely those of my friend. Or selection may act on the memes indirectly; I die, and all my memes disappear. In this latter case, exosomatic storage may have some impact because, of course, stored memes survive while we die. However, I remain sceptical, since if the memes are strong enough to exert a force when stored exosomatically, they would likely have survived the selection process when their originators and possessors were alive and would have been transmitted to others. In other words, 'good' and 'fit' memes don't need exosomatic storage. On the other hand, cultural selectors may change in time, and when they do exosomatic storage may allow for the selection and propagation of ideas, memes, that are 'before their time'. We shall see a useful scientific example of this below.

A transmission device, usually a form of language, moves the selected memes about in space and conserves them in time. This combination of differential selection and transmission leads to differential conservation of memes over time, and hence results in changes in meme frequencies in the cultural meme pool in time. Furthermore, memes are not immutable. They change, for what-ever reasons, and these changed forms are further subjected to selection, which leads to further changes in the constitution of the meme pool. The result is descent with modification; and what is being modified is the culture itself. All this adds up to cultural change being the result of cultural evolution.

While this may sound a plausible account, if it is to be maintained then many details will have to be filled in and worked out. For one thing, if cultures are indeed evolving systems, then, in the language of universal Darwinism, the replicators, interactors and lineages must be identified and their properties understood. As a beginning one can assert that memes are ideas, but what exactly *is* an idea? How many are there? If an idea can be captured by a brief sentence, then, remember, each of us has potentially some 10^{30} memes in our heads. But would these all be memes? And can there be nonsense memes (like a gruwidget), or memes for the non-existent in the sense that they refer to things that do not actually exist in the world (such as a hairy bath-tub or mythical beasts like unicorns)?

Furthermore, if ideas correspond to brain states, then which brain states? In an organ made up of millions of highly interconnected nerve cells, the number of possible 'states' within even a restricted nerve sheet of about 100,000 cells (which is a very modest affair as brains go) is astronomical. And we just do not know what small changes of state within such a nerve sheet mean psychologically. It is obvious that every day is different from every other day for any living thing, and especially so in an organ of the complexity of the brain. In that most general sense, the meaning of 'market forces' must be placed within the slightly different perspective of background experience each day; however, as already pointed out, the core meaning of 'market forces' may somehow be retained in a quite stable form. So exactly how we define a meme is uncertain. A 'belief', defined as a relatively fixed core meaning, tolerant to minor changes about a fuzzy edge, might be a more appropriate and workable image of what a meme is; and a change in belief might be a more credible notion of what a change in a meme is.

An approach related to the notion of meme as belief is that of meme as a 'bundle' of ideas, some kind of higher-order knowledge structure rather than just a simple memory. For example, my restaurant meme, which I share with almost all other members of my culture, is that a restaurant is a place where one goes when one is hungry, that it is a place where one normally eats sitting at tables, that people will prepare and bring the food to one and that one must pay for it all. Contrast this network of knowledge, sometimes referred to in the technical literature as a schema, with the much more specific and elemental meme that Joe's Greasy Spoon is located at a specific place, that it has pink flocked wallpaper and is run by a man called Bill. Of course, both the specific memory and the schema might be memes. But it does seem more credible that culture is importantly made up of higher-order aggregates of knowledge, and that it is these that allow us access to, and the ability to function within, a culture.

But is it reasonable to think of ideas, memories, brain states, as replicators at all? The answer, I think, is yes. Because the brain is such a complex and dynamic organ whose overall state is in constant flux and which has an ever-varying state of functional connectivity caused by different patterns of cells firing in time, it is not just plausible but essential to think of it as being able to re-constitute, re-member and re-plicate specific states. Without such an ability the brain could never conserve information. As an example of what I mean, ask yourself where you were on New Year's Eve ten years ago. This memory is gained by a process of reconstruction, a kind of protracted problem-solving activity. You may begin by establishing what country and what town you lived in a decade ago; then, perhaps, whom you were living with; then what job you had at the time. This may be combined with quite a bit of backtracking, such as, 'Last year I was at home in London, but the year before that I was away on a business trip, and the year before . . .', and so on. Eventually, after quite a long series of such mini-problem-solving tasks, you will arrive at a specific memory, a brain state, corresponding to the name of the place where you were ten years ago. And it is likely that the brain state correspond-ing to that place name, say Val d'Isère, is roughly the same as the brain state corresponding to Val d'Isère yesterday or last year or last decade. Now, you have not been walking around with a specific brain state corresponding to Val d'Isère in your head all this time. That particular brain state has not been maintained constantly. So the memory of the name, or the idea, of Val d'Isère is a re-membering, a re-plication of a brain state.

In the same way, 'market forces' is not an idea that is constantly with me, or anyone for that matter, not even economists of a certain kind. Few ideas *are* constantly present for anyone. Asked to say what 'market forces' means, I will have to re-member the bits and pieces that go to make up the idea: 'Not an actual thing . . . an abstraction . . . it concerns the effects of demand on availability . . .

availability determines price', and so on. As with Val d'Isère, so with a notion like market forces, a brain state can be reconstituted when such a state was not actually present all the time in the past. This characteristic of neural replication, of making a copy, is different from the replication of genes. In the latter case, a specific chemical structure is constantly present and, at certain points in the life cycle of the interactors carrying those genes, copies, literal copies, are made of the 'original'. Clearly neural replication is different. Here copies of brain states present some time in the past are reconstituted, such structures being only potentially present, not actually so, in the intervening time. How such actual brain states are re-established from potential brain states is, for the present, a process whose details are entirely unknown by brain scientists. We simply do not know how it works.

Then there are the psychological complications of cultural replication. Over half a century ago F.C. Bartlett, a British psychologist, conducted a series of experiments that are fundamental to the understanding of human memory. Instead of studying memory for single words or single events, Bartlett studied memory for stories. People were exposed to passages of prose, fragments of stories, and at some later time asked to recall the stories. What he found was that memory for narrative is an active and changing form of memory, subject to much influence by the previous and intervening experiences of the individual. Especially relevant to understanding culture were Bartlett's experiments on chains of story-tellers in which he studied the systematic alterations to the contents of a story which is passed from person to person, each listener becoming in turn the story-teller for the next. This looks close to an experiment tailor-made for those interested in culture. Bartlett described several forces that shape the changes in stories across story-telling episodes. One is for stories to become compressed and compacted down to rather bare essentials. Though startling or amusing snippets would be retained, the general tendency is to retain meaning but eliminate, or fabricate, details. Another force for change he called conventionalization, peculiarities of style and opinion being lost to an increasingly bland version of the original. Thirdly, Bartlett found that stories become 'rationalized', by which he meant that curious or mysterious assertive content slowly changes into a meaning acceptable and intelligible to the average listener. So feed a rich, clever and slightly obscure story into a group of listeners, and what comes out at the end is a rather colourless and none-too-accurate husk of a tale.

Bartlett's findings on just how changeable narrative memory is have been subsequently confirmed and expanded upon by many studies of eyewitness testimony. The gloomy message that such work has for our understanding of cultural change – namely that it is very, very complicated – might be tempered, again, by considering the effects of exosomatic storage. If psychological studies indicate that word-of-mouth cultures must be characterized by instability on the one hand or dullness on the other, then exosomatic storage of texts may explain how subtlety, richness and beauty can be retained in the contents of many cultures over long periods of time – witness the Bible or the Koran. So it just may be that cultures that have existed without exosomatic storage of memes have been and are dull by comparison with cultures that have the written word. In this, in a sense trivial, manner, exosomatic storage may have a more important role to play in the evolution of cultures than was earlier suggested.

There is one really important difference between genes and memes, a difference that may have profound effects on the way that cultures evolve as opposed to how biological systems evolve. This concerns the way in which memes are inherited. It will be remembered from Chapter 2 that prior to Mendel it was believed that genetic inheritance is a result of the blending of characteristics from each parent due to the blending of whatever it was that offspring inherited. The children of a tall mother and a short father, it was believed, would be somewhere in between, because the inherited material that determines height would meld and blend. The observation here of what happens to the phenotypic character of height is correct, but the reason is not because genes have blended, but because height is caused by many genes and developmental factors, so the end result of inheriting so many genes at random is indeed a kind of middling one. This, however, is not blending of what is inherited. Blending is typified by the expectation that a child of blue- and brown-eyed parents would have some muddy in-between eye colour which that child would then pass on to its own offspring because the genes determining eye colour had blended. Blue or brown as an eye colour would give way for all time to a muddy blend. But, of course, the children of such parents do not have a blend of parental eye colour. They have either blue or brown eyes.

Mendel's work demonstrated what should always have been obvious to even casual observers of traits like eye colour, namely that genes do not blend but retain their integrity even when they are not expressed by an individual, and their effects are passed on undiluted to offspring who may express a gene that their parents did not. Genes do not blend even if the characters that they determine do so.

Memes, however, may blend. It is at least reasonable to consider that cultural inheritance may, under some circumstances, result in blending, especially as, unlike genes in sexual reproduction, memes may be inherited from multiple sources. One hears about God within one's family as well as at school; religious belief is debated in the media; and people discuss religion with their friends and acquaintances. Once again we do not know the answers to this problem with any certainty, but the multiple parenting of culturally transmitted religious ideas may well lead to some middling form of belief; or it might happen that such belief may be overwhelmingly influenced by some single important 'parent', some special teacher. It is not possible yet to come to any general conclusion about blending inheritance of memes and the effects of multiple parenting, but people who work in this area are aware that such possibilities have to be taken into account and that cultural evolution might have certain special characteristics that are absent from biological evolution.

Remember Dawkins's characterization of replicators by the trinity of qualities: longevity, fidelity and fecundity. Remember also the comments in Chapter 3 on why conventional evolutionists have so valued longevity in their choice of unit of selection – because adaptations must be retained as stable parts of phenotypes over many generations if they are to have evolutionary consequences. Culture seems to be different, though how different is uncertain, because some memes do have remarkable stability and others are short-lived. If cultures do change because they evolve by the processes of universal Darwinism, then we must have a theory and an understanding of culture that take into account both the potential rapidity of change in cultures and some degree of stability. In general, though, the cultural unit of selection may be much less long-lived than a gene. Indeed, all of memetic replication looks different from genetic replication: not much longevity except for core conventional meaning and startling detail; very little fidelity apart from simple memes; and a fecundity that probably varies from person to person as a result of differences in cognitive capacity yet to be understood.

What of cultural interactors? What are they? One possible interpretation is that it is behaviour based upon memes that selection acts upon. Having acquired the meme for market forces, for example, I then set up in business, and as a result of this action I adjust my idea of what market forces are and how they act in the light of my experience. In this view, then, behaviour is the interactor. Another approach is that in argument and discussion, as in science, interactors may take the form of reasoned exposition or application of conceptual theory. I may, for example, think through the issue of market forces and come up with an altered meme; perhaps I might come to think of them as having additional properties such as inducing poverty and restricting choice rather than being a creative economic force, or I might have started from a position hostile to the notion of market forces and then altered the meme as I pondered the parlous state of countries that do not have market forces.

Another feature of cultural evolution that has yet to be properly understood is the matter of just what the selectors are. Consider the matter of papal celibacy. Biologically a pope would be considered as having near-zero fitness – he will leave no direct offspring bearing his genes. Yet the pope is a figure of enormous cultural significance: his cultural fitness is very high indeed, he being instrumental in the propagation of memes in a manner far more effective than ordinary people are capable of. This schism between biological and cultural fitness can only be understood by invoking culture-level selectors that are largely, but perhaps not wholly, decoupled from biological-level selectors. Culture is evolutionary causation operating from within the brain writ large.

What exactly such culture-level selection is, how it comes about and what it is connected with (especially the matter of whether and how it is connected to primary and secondary heuristic selectors) also remain unknown. Science, as has been said on a number of occasions in this book, is a special form of culture that is transformed in time by evolutionary processes. Because of the nature of the enterprise it is far better documented than, say, the rise and spread of religions. Yet it is not much better understood.

A famous example among historians and philosophers of science is what are known as plate tectonics and continental drift. It is now accepted as the centrepiece of modern geophysics that the continents slowly, ever so slowly, ride about the surface of the Earth on massive plates that drift here and there. When this idea, this meme, was first put forward in 1915 by the German geophysicist Wegener, it was almost unanimously rejected by that particular scientific subculture. Since the 1960s it has stood as the central conceptual plank of geophysics. There are many possible reasons why a meme that was selected against at one time was then so strongly selected for at another, and the details need not concern us here (though it might be noted that the plausibility of the idea of the continents sailing gently across the surface of the planet gained credence when mechanisms explaining how it might be happening were discovered – testimony to the power of mechanism as scientific explanation). But what is clear is that, in that half-century between the initial postulation of continental drift and the subsequent acceptance of plate tectonic theory, the individual intelligences of scientists had not changed, and neither had the biological conditions for the evolution of *Homo sapiens*. What had altered was something at the level of that particular science. (This is the example of exosomatic storage referred to earlier: it is an excellent demonstration of the possible importance of such storage, whereby a meme is saved from being selected out

of a culture at one time, and then selected for at a later time.)
Roughly the same kind of thinking applies to fashions in dress,
where it is patently obvious that forces of selection operate at the
level of the fashion industry itself. And yet the possible effects on
cultural evolution of selection processes operating at the level of
individual intelligence, as demonstrated by Bartlett's experiments,
should not be forgotten or ignored.

Cultural lineages are written into our history, and once we
know better how to identify cultural replicators and interactors
then we will be better able to disentangle these lineages. But again
there is a likely disanalogy with biological evolution. It may be
that cultural lineages themselves have become, or will become,
memes. If this is so, then cultural evolution really does have an
extraordinary dynamic quality that is absent from biological evolution. However, it may also be that the conception corresponding
to a lineage cannot be a meme, and this would indicate just how
much conceptual work needs to be done in sorting out the basic
units of cultural evolution.

Now, the preceding paragraphs have all been about mechanism.
But why has culture evolved at all? And how should it be
understood in its relationship to biological evolution (the primary
heuristic) and psychological evolution (the secondary heuristic)?
Well, it could be that culture has not evolved at all, in the sense
that it did not evolve for the reasons of whatever function it now
serves; it may merely be a consequence of the evolution of
language in combination with the presence of various other human
intelligences. In other words, culture may just be a manifestation
of the human secondary heuristic rather than being some entity in
its own right. An alternative view is that culture has evolved as a
specific human attribute, the reason for its evolution being its
capacity to supply behavioural adaptations to complex, changing
environments even more rapidly and effectively than intelligence
acting in the lone individual. After all, learning about some feature
of the world, or how to perform some skilled act, through the
medium of some other individual teaching the learner may often
be a much more rapid way of gaining knowledge than learning
through one's own experience. It took a long time to invent the

wheel. Once invented, its advantages and uses could be rapidly taught and expanded upon.

There is another way of putting this. Perhaps the central characteristic of humans that makes us human is our extraordinary capacity for detecting and exploiting short-term stabilities in the world. Sensitive to the existence of so many short-term stabilities in our environment, we have a great deal to learn – more than any other species. We also have an unusually long period of infantile and juvenile dependency upon adults. Could it be that culture evolved as a way of exploiting the long dependency of our young in order to solve the problem of mastering some of those short-term stabilities by vicarious experience and the sharing of knowledge?

If the latter view is correct, then it may be helpful to think of culture as being a third-level heuristic, another form of Darwinian machine and hence another means of gaining knowledge of the world based upon evolutionary processes, those same processes that operate for the primary and secondary heuristics but involving separate mechanisms. In this case the complete system of human knowledge would be a three-level control hierarchy, each level operating on evolutionary principles, each with its own units of selection, each comprising subsystems of replicators, interactors and lineages, and each operating at successively higher frequencies and hence each sensitive to ever-more-fleeting short-term stabilities. The whole image is of a hierarchy which has levels oscillating within different frequency ranges embedded within a world comprising change which is occurring within different frequency ranges. In this image, human knowledge is a web of relationships that matches different levels of the hierarchy with different features of the world as a function of the matching of the rates of oscillation of the levels of the hierarchy with the frequency of change of different features of the world.

There is one final question that our discussion of culture raises. The secondary heuristic is nested under the primary heuristic, and the consequence of this is that human intelligence specifically, and all animal intelligence in general, is constrained by and acts in the service of its primary heuristic. Does such nesting apply in the case of the tertiary heuristic? On the one hand, the evidence would seem to count against such a view. Papal celibacy, weapons of mass destruction and the devastation of the natural world and its resources seem to point to cultural forces that are quite independent of the primary heuristic. If the perpetuation of genes, the primary heuristic replicators, is what the primary and secondary levels are ultimately about, then many of the consequences of culture seem to run counter to this general goal. I think, though, that this view is wrong, a kind of conceptual illusion based upon our present incomplete understanding of culture. Some of the seemingly 'unnatural' features of culture, like sanctions against sexual behaviour or the establishment of certain kinds of cultural institutions, may be culture-level manifestations or consequences of cultural selection set by the primary and secondary heuristics. Napoleon, for example, when asked why he had restored the church in France replied that he saw the greatness of the church not in the promise it held out for individual salvation but as a force for social cohesion. Could this be an instance of the lower-level heuristics somehow reaching up to the third level but in ways, and by routes, that we do not yet understand? Far-fetched as this may seem, it is not a possibility that we can rule out.

Other 'unnatural' consequences of culture, like the depletion of the ozone layer or the greenhouse effect, are indeed unnatural and deeply damaging to the primary and secondary heuristics. But we know this already, and that is the point. Not only have these effects not passed unnoticed, but they have been detected by the tertiary heuristic of science itself. The total human knowledge system is well aware of the potential destruction of our species by the activity of the third heuristic, and we are beginning to do something about it. If human culture were truly detached from human biology, then alarm bells would not now be ringing. We would blithely destroy the world that we evolved upon and then take happily to living science-fiction lives as some kind of semi-robotic creatures drifting through the universe in dreary space vehicles. That may be the future, but I doubt it. We will use the tertiary heuristic to maintain the Earth in a state fit for the primary and secondary heuristics to operate in.

More than any other part of this book, this section has sung constant refrains of 'we don't yet know' and 'it is not yet understood'. This is simply because culture is the most complex thing on earth. It has its origins in, and ultimately we will understand it by way of, genetics, development, neuroscience and cognitive science, ecology, ethology and anthropology. We will, though, surely succeed in understanding it one day, because culture is a very natural phenomenon.

SUGGESTED READING

Barkow, J.H., Cosmides, L. and Tooby, J. (eds.) (1992) *The Adapted Mind: Evolutionary Psychology and the Generation of Culture*, New York: Oxford University Press.

Fodor, J.A. (1983) *The Modularity of Mind*, Cambridge, Mass.: MIT Press. A famous and admirably brief account of the structure of mind.

Gleitman, H. (1991) *Psychology*, 3rd edn, New York: Norton. Perhaps the best introductory text in general psychology available.

The Philosophical Problems in Perspective

═══

Knowledge as commonly understood is a special kind of adaptation. And all adaptations are knowledge. The fleshy water-conserving cactus stem constitutes a form of knowledge of the scarcity of water in the world of the cactus, and the elongated slender beak of the humming-bird is a manifestation of the knowledge of the structure of the flowers from which the bird draws nectar. In both cases it is a very partial and incomplete knowledge, but knowledge it is. Now, does this claim that a cactus or a bird has knowledge of the world not degrade either the ordinary meaning of the word 'knowledge' or its more elevated philosophical reference?

I think not. First, remember that to say that a bird has knowledge when one is referring to a specific characteristic of that bird like its beak or feathers is really to use a form of shorthand. What is actually meant is that knowledge is a complex set of relationships between genes and past selection pressures, between genetically guided developmental pathways and the conditions under which development occurs, and between a part of the consequent phenotypic organization and specific features of environmental order. As long as the genetical and developmental components of this rather large set of relationships are always taken as read, it suffices to point only to the expression of knowledge in terms of phenotypic organization and environmental order.

What applies to the beak of a bird or the stem of a cactus plant applies also to knowledge as commonly understood. To take a trivial but simple example, I know who won the 1990 World Cup football competition. The Federal Republic of Germany, as it then was, is a specific feature of environmental order, and my brain state conforming to Germany as the winner of that cup is a part of the internal organization of my phenotype. As it is with all forms of knowledge, so it is with all such instances of human knowledge which we normally locate somewhere between our ears. This particular instance of knowledge, namely that Germany won the World Cup, is really only the visible, or potentially visible, part of a complex multiple-layered and historically ordered hierarchical structure involving the genes which code for the brain structures that enable me to gain knowledge, development which led to the establishment of the required brain mechanisms, brain and cognitive states that are the present embodiment of that knowledge, and culture and its artefacts that allow me to learn rapidly and accurately what is occurring in a distant part of Europe.

We employ our capacity to know something of our world in a relatively trivial, non-life-or-death way much of the time: for example, I also know that England, alas, have not won the World Cup since 1966, and that Buzz Aldrin was the name of a man who went to the moon. This should not detract from what is a complicated and rather grand control hierarchy that extends across a great range of time from the thousands if not millions of years that are required for genetical changes to the millisecond durations by which we measure the changes in state of the brain. Lacking a brain of any kind, the cactus has its knowledge built upon a less complex structure of genes and development. But otherwise, big as that otherwise is, we, the cactus and the bird have much in common. It is in the understanding of what we have in common that we gain deep insight into the nature of human knowledge, and one therefore cannot reasonably think of either that commonality or the claim that birds and cacti too have knowledge of their world as being somehow demeaning to our understanding of the word 'knowledge'.

So it is merely a human conceit to think that knowledge is something that is both unique to our species and located only in our heads. Knowledge is a pervasive characteristic of all of life. It exists in all adaptations in all living creatures. Seeing knowledge in this light is far from demeaning. It gives us a deeper, more fundamental understanding of what knowledge is and what a scientist means when the word 'knowledge' is used. It casts knowledge as a natural phenomenon which is amenable to scientific analysis and understanding.

But does the science of knowledge outlined in the previous chapters have anything to offer traditional analytical philosophy in its quest to further the understanding of knowledge, specifically the classical problems posed by epistemology, some of which are outlined in Chapter 1? Well, both yes and no, and it is with a brief explanation of this Janus-like answer that I want to conclude this survey of an evolutionary scientist's eye view of the problem and nature of knowledge.

WHAT SCIENCE CANNOT TELL PHILOSOPHY

Philosophy is concerned with an understanding of the world gained through thought, and 'the logical clarification of thoughts', to quote Ludwig Wittgenstein. Excellence of argument, coherence of argument, lack of contradiction, these are what the philosopher strives for, whether it is ethics, epistemology or some other branch of philosophy that is the object of study. What philosophy does not require is empirical validation of its findings – indeed, to expect this is to fail to understand what philosophy is about. So philosophers do no experiments; and their observations of the world are illustrative and informative rather than being absolutely determining, the final judge, as is the case in science. If philosophers did experiments, then, of course, they would be scientists. This independence from the empirical activity that is so crucial to science – only a half of science, that is true, but a half is a great deal – means that philosophers are entirely justified in distancing themselves from evolutionary epistemology *as science*.

Everything in Chapters 2 to 6 of this book is either what science knows with varying degrees of certainty, because it is the product of studies that have already been conducted, or cast in such a way that it is, or one day will be, amenable to scientific study. While it might be judged that philosophy ignores science at its peril, none the less the nature of the enterprise allows it to be very fussy about what science it does take notice of. Facts as facts are seldom of any intrinsic interest to scientist or philosopher. In science facts become interesting only when they form patterns, that is, when they begin to conform to theories of some kind. And scientific facts are only of interest to philosophers if they speak to issues of philosophical importance. If they do not, then philosophers can quite properly ignore them.

A control hierarchy of mechanisms for gaining, storing, expressing and transmitting knowledge is a scientific view. I will shortly argue that there are several features of the scientific theory of knowledge that certainly should not be ignored by philosophy. But in one respect at least evolutionary epistemology is silent about an issue that is central to the philosopher's approach to epistemology. This is the problem of justification, of believing that a knowledge claim is reasonably well founded, of being able to assert that one is certain in knowing that one claim is true and another is false.

The reason for evolutionary epistemology's inability to pronounce on this issue is as follows. We can ask whether organisms survive and reproduce because their adaptations, like water-retaining properties in plants or beak shape in birds, are relatively true reflections of the world that they have to interact with and exploit. The answer is a cautious yes. The caution tempers any absolute pronouncement, because all organisms are bundles of organized features, some of which are good adaptations, some of which are not very good adaptations, and others are not adaptations at all. As we saw in Chapter 2, fitness, in the sense of reproductive competence, is, in the words of the great American evolutionist G.C. Williams, a statistical abstraction. *All* the adaptations and all the non-adaptive phenotypic features of an organism sum together algebraically, and if the characteristics of positive value outweigh those of negative value, then that is a creature more likely to survive and reproduce than one where the situation is reversed. But no creature has only positively weighted adaptations of high value. At some time in the history of a species all of its members will possess negative attributes, and some of these will get through the filter of natural selection by riding on the backs of attributes of high positive value. Eventually these negatively valued attributes and those of low positive value will be selected out. And, of course, life does not stand still: novel attributes, often of less than excellent value to the fitness of the individual organisms in a population, are constantly appearing, changing and disappearing.

What this means is that one can take an organism, any organism, and point to this and that feature. Most of those that are adaptations – and remember, not all will be – will be competent adaptations, by which I mean they will comprise reasonably good knowledge and so contribute to fitness. But not all. Hence the caution accompanying the affirmative answer to the question as to whether organisms do indeed survive and reproduce because their adaptations are relatively good reflections, are relatively good knowledge, of their world. One may not always be certain as to whether in any one case survival is contributed to by knowledge or whether we are just fooled into thinking that this particular organism has good knowledge because it survives. However, put crudely and simply, it is generally the case that knowledge furnished by the primary heuristic is usually well justified, if only because it has been a very long time in the getting, with selection repeatedly scrutinizing and improving its accuracy.

But when philosophers ask about epistemological justification they are not talking about enzyme structures in bugs or the skin coloration of the chameleon. They are asking how you or I can be certain that we know that it was Germany and not Italy that won the 1990 World Cup, or that it is or is not raining today. How can we tell that one human belief or knowledge claim is false and another is true? In other words, philosophers are concerned with what humans claim to know in the commonly understood sense of knowing.

Now, knowledge in this ordinary sense is a product either of individual intelligence, the secondary heuristic, or of culture, the tertiary heuristic; and if the reader recalls the gist of Chapters 5 and 6, it is that these subsidiary heuristics evolved precisely because the primary heuristic could not know with any sort of precision about states of the world whose frequencies of change exceed those of the operating characteristics of the processes and mechanisms of the primary heuristic. That might lead one to expect that the products of the second and third heuristics would be accurate and true. But, just as the adaptations generated by the primary heuristic are of varying degrees of truth and may even be untrue (though in the fullness of time they will tend towards the truth), so it is too with the knowledge that results from the operation of culture and individual intelligence. To be sure, the primary heuristic does what it can to ensure that what we come to know by the secondary heuristic is roughly in the right 'ball park' – hence the constant insistence on the incorrectness of *tabula rasa* views of intelligence. But the priming by the primary heuristic will never guarantee good knowledge at all times. More to the point, having 'incorrect' or false knowledge, even a great deal of false knowledge, does not mean that the individual will fail to survive and reproduce. Of course, it may mean just that, which is why the primary heuristic itself is constantly evolving more accurate priming in the operation of the secondary heuristic. Also, the secondary heuristic itself has a whole host of devices by which it can and does correct its knowledge, false beliefs being substituted by less false or more true ones. But whatever the efforts of the primary heuristic to get the secondary heuristic to walk the path of truth, and however full of self-correcting mechanisms intelligence might be, the secondary heuristic does make errors, and so we do all come to have widely incorrect or plain untrue beliefs in varying numbers and hold them for varying periods of our lives.

The tertiary heuristic, even if nested under the primary and secondary heuristics, is subject to even less control by the more-often-than-not-true-belief-generating primary heuristic and the quite-often-untrue-belief-deriving secondary heuristic. The net result is that cultural beliefs are even more often untrue or widely inaccurate. As a general rule, though with one important exception to which we will return in the final paragraphs of this book, the further we climb away from the primary heuristic, the more likely it is that the knowledge that is generated will depart from the truth.

Human knowledge as commonly understood, then, is quite

often untrue. Whether it is or is not more often true than untrue, closer to the truth than further from it, is irrelevant. Even if it were only very rarely untrue, but the holders of the untruths survived and reproduced, that would be enough to nullify any foolish claim by evolutionary epistemology to overcoming the justification problem. Our knowledge is fallible, and even though we understand why, which is itself surely a real achievement in knowledge, that understanding itself does not help the traditional analytical philosopher. Only if survival and reproduction are absolutely correlated with knowledge could they be an infallible guide to true belief. But since this is not the case, evolutionary epistemology has nothing to say about the knowledge justification problem.

That, however, does not mean the end of the epistemological world. For one thing, the justification problem may simply be unsolvable – in effect a non-problem that philosophers have worried at for thousands of years to no avail. For another, the science of knowledge certainly does have some interesting and informative things to tell philosophy about other epistemological issues, and it is to these that we can now finally turn.

WHAT SCIENCE CAN TELL PHILOSOPHY

The etymological and common-sense distinction of knowing 'by the senses' and knowing 'by the mind' which was considered in Chapter 1 is strongly supported by the science of knowledge. Moths that are attracted to light approach light sources because of the way that their sensory systems are wired up to their effector systems – hard-wired, in the jargon of the neurosciences. This wiring is present in the adult moth when it emerges from the pupa stage and it never alters. So when removed from the light the behaviour of the moth is not changed by the experience of light; and when subsequently re-exposed to a light source its behaviour is again unaffected by the prior light experience. In other words, moths do not undergo any enduring changes in brain state that last beyond the immediate neural effects of light exposure. This is all a rather tortured way of saying that moths have no memory of the light or of the effects of flying towards it. Not having any ability to form individual memories, the moth can only behave in ways appropriate to current experience. The moth can only know by the senses; to know by the mind requires, at a minimum, a memory, and is often accompanied by the ability to manipulate those memories through the processes of thought.

The distinction, then, between knowing by the senses and knowing by the mind maps neatly on to knowledge gained by the primary heuristic and knowledge gained by the secondary and tertiary heuristics. All instinctive behaviour patterns are examples of knowing by the senses, because intelligence is never a cause of such behaviour patterns. The knowledge that underlies instincts is gained and stored wholly in the gene pool and expressed during individual development as a particular part of the hard-wiring of the nervous system, which, when activated, gives rise to the instinctive behaviour in question. Konrad Lorenz used to refer to such knowledge as cognition without memory. By contrast, when behaviour is a product of intelligence or culture, a part-cause of that behaviour is always a brain state, an individual memory, which allows for knowledge claims, or gives rise to behaviour which would be based on a knowledge claim if the creature could talk, or gives rise to behaviour 'as if' the animal knew about events in the past. And if the memory is manipulable by processes of thought, behaviour might also embrace knowledge claims about future events, that is, predictions. So knowledge involving the secondary or the tertiary heuristic is always a matter of knowing by the mind, where the knowledge is about events distant in time from when the knowledge claim was made, or from when the behaviour implying a knowledge claim occurred. The distinction is well grounded within an evolutionary epistemological approach to the science of knowledge.

Now, the common-sense distinction between knowing by the senses and knowing by the mind certainly does not fit neatly into the philosophical distinction between rationalism and empiricism. But there is a certain resonance which allows us to consider what evolutionary epistemology tells us about the relative merits of these two ancient schools of epistemological thought. It will be remembered from the first chapter that the distinction turns on the weight given to sensory experience.

The rationalists asserted that the world could not be accurately known through the sense data, if known at all in this way. The senses were, for Plato, a destructive distraction from the truth. When we come into the world, according to Plato, we already have real knowledge, truth, within us, but have to work hard intellectually, through the study of logic, mathematics, and the like, to realize that with which we are all born. For Plato, 'knowledge is not perception but a kind of reminiscent vision', in Bertrand Russell's lovely phrase. So, for rationalists like Plato and Descartes, knowledge was very much 'in the mind'.

The empiricists, by contrast, laid varying amounts of weight on the importance of sensory experience, generally believing that it is only through the senses that we can, if we can at all, come to know something of our world. However, they certainly did not deny the existence of a knowing by the mind in terms of memory and thought. Perhaps the best way of phrasing the empiricist position is to say that we come to know by the mind, but we do so by way of a route which always begins with knowing by the senses. Of course, a further distinction between empiricists and rationalists lies in the empiricists generally denying that we come into the world with any knowledge at all, while innate knowledge was an absolute requirement of the rationalist position.

The particular evolutionary epistemological view argued for in previous chapters holds that both positions are correct in some ways and incorrect in others. Both have part of the answer to the problem of knowledge, but lacking the historical perspective of evolution and the notion of adaptation, neither could ever have been an adequate account. They simply lacked the conceptual fire-power needed to crack the problem. The answers only begin to come when one considers creatures other than humans.

Taking all of life into account, and not just humans, knowledge certainly did begin with sensory experience. Intellect, the secondary heuristic, did not come into existence for thousands of millions of years, during which time life certainly did have knowledge of the world, all of it being gained by the primary heuristic in which experience of the world shaped the form and function of living things through genetic and developmental mechanisms. Yet right from the beginning, in a wholly 'empiricist world', living creatures came into this world with innate knowledge, but it was knowledge that was entirely without intellectual basis. It was only as the world became more complicated with more rapid rates of change that a small number of life's forms evolved secondary heuristics as a means of combating an increasing problem of uncertain futures. But this intrusion of rationality into an already existing ocean of knowledge, despite the astonishing power with which it sometimes manifested itself – Plato's mind, for instance – occurred because of the limitations of sensory knowledge. None the less, it was rooted in these sensory experiences and in sensory knowledge.

However, in one crucial respect the rationalists were absolutely right. Much of knowledge, perhaps all of it, including human knowledge as commonly understood, is innate, or based on innate structures. We all come into the world knowing what it is we have to learn and think about. The rest of our intellectual lives are spent filling in the spaces.

Science, then, tells us that the rationalists and empiricists were both right and both wrong. Here we have an epistemological issue of no small weight that is resolved by modern science. Evolutionary epistemology also gives us an interesting perspective on the ideas of the person that many consider the greatest philosopher to have lived since Plato: Immanuel Kant. Once again I blush at the unseemly brevity and haste with which a part of a great philosophical position is presented.

Kant was certainly no thoroughgoing rationalist, in that he did not believe that reasoning alone could lead to a knowledge of the true nature of things. Furthermore, while he did not believe that we can know everything of the world just through our senses, if we can know anything of it at all, he certainly did believe that sense data are caused by something out there in the world. And appalled by the scepticism of Hume and the power of Hume's arguments, Kant looked for another way of making sense of experience. The problem, as Kant saw it, was that the ironical outcome of the potent criticisms made by the empiricists of the possibility of knowledge being based upon sensory experience might be an extreme form of subjectivism. It is not too far a step from Locke's judgement that 'Since the mind, in all its thoughts and reasonings, hath no other immediate object but its own ideas, which it alone does or can contemplate, it is evident that our knowledge is only conversant about them', to some form of solipsism in which it is only myself that I can ever know, while the external world has no existence independent of myself.

Believing that reason alone is not enough and that sensory experience is indeed the result of something in the outside world, Kant held that knowledge comes through a conjunction of some internal set of factors and sensory data. In this sense if no other, he was neither rationalist nor empiricist. We come to know the truth of certain things because, he argued, it is only through their truth that experience of the world is possible. Needless to say, Kant's arguments are complex, but one relatively simple entry into his thoughts is by way of certain distinctions that are central to his position. One such distinction is between analytic and synthetic propositions. An analytic proposition is self-contained and non-contradictory; reason alone will tell us whether such a proposition is true or false. By contrast, a synthetic proposition is based upon experience and cannot be arrived at by reasoning alone. For example, the propositions that it usually rains on weekends in Britain and that the line that you join in a supermarket always ends up being the one that moves most slowly cannot be arrived at by analytical argument, but only by experience. If you want to know about weather patterns, don't ask a logician, talk to a tennis player.

The other distinction that Kant drew was between empirical propositions, which are about things that can only be known through sense data (you have to get out there and measure the length of the lines in the supermarket), and a priori propositions, which are arrived at outside of individual experience, that is, which involve innate causes. The trick that Kant used was to claim that certain things are known both as synthetic and a priori.

Mathematics is one such; causality is another. One has to experience relationships such as greater than or smaller than and to learn that three pieces of cake are half the number of six pieces, but we are innately capable of profiting by such experience and becoming numerate. In the same way, we need to experience the conjunction of two events in order to claim that the one causes the other, but the capacity to appreciate causation is a priori.

But how can something be due both to experience and yet also to some knowledge that is prior to experience? The careful reader of the last two chapters will have no problems with the answer, so it is worth remembering that Kant was writing and thinking over 200 years ago. His answer was that we come into the world with certain 'intuitions' (about space and time) and certain 'categories' (twelve to be exact, which he grouped under the headings of quantity, quality, relation and modality). The intuitions and categories are a kind of inborn set of spectacles through which we see the world and by which we think that we come to know it. More than that, for Kant coherent experience of any kind would be impossible without these spectacles. Experience plays upon these innate intuitions and categories, the result of which is what we call knowledge of the world. But – and for Kant it was a very big but – while there are things 'out there' (which he called 'things-in-themselves' or 'noumena'), which in some way cause our sensations, because we interpret these through the 'ideal' categories and intuitions of the a priori, and because the ideal categories and intuitions do not necessarily represent the ordering of the world as it really is, the inborn spectacles prevent us from ever really knowing what the world is like. The a priori stands, as it were, between the world as it really is and our experience of it. So we can never actually know the world; we think and act as if we do, and there certainly are things-in-themselves out there, but our knowledge is always 'ideal' or 'transcendental', never real. So it is that the a priori categories and intuitions are what makes experience possible – that conjunction between the inborn and the data from the senses. But the a priori also stands between us, the experiencers, and the things-in-themselves, which can never be known.

Now, Kant's transcendental idealism is, to the late-twentieth-century scientist, strangely at variance with the scientific realism that washes through our thinking. All of modern science is based upon the assumption that what is out there can most certainly be known, and known better and better as science progresses. What we know is a reflection of the things-in-themselves. Furthermore, the subjectivism of Kant's intuitions doesn't tell us why we consistently see the world in certain ways. 'Why, for instance, do I always see people's eyes above their mouths and not below them?' asked Russell. According to Kant, something like eyes and mouths do have an existence, but there is nothing about Kant's spatial intuitions which would have us see them in some consistent spatial relationship to each other. Modern scientific realism, by contrast, assumes that the space and time of our perceptions do have counterparts in the world outside. Mouths *are* always beneath eyes, and what we see corresponds at least partially to what is out there. Our sensory data may not be perfect, they may not be complete, and they may not be direct. But the data do bear an element of correspondence to the things-in-themselves.

There is another reason for rejecting Kant's thesis of the essential unknowability of the things-in-themselves. Living creatures survive by exploiting energy sources in the world outside of themselves, and by avoiding events or entities that would destroy them. They do not do this by chance, and the three and a half billion years for which life has managed to succeed in this enterprise have not been some kind of enormous fantasy. The success is a result of living things being able successfully to match their own organization to the order of the outside world. That is, they can and do indeed know about the world outside of themselves. Inaccurate, incomplete and partial though the knowledge might be, it none the less is a practical working knowledge. Our lives depend on it. That we survive at all is proof that knowledge is possible. To paraphrase a comment once made in a rather similar context, God would no more have given us brains by which we could not know the world than he would have given us eyes by which we could not see.

This insistence on a claim to some form of knowledge, workably good knowledge, does not contradict the assertion of the previous section of this chapter which pointed to the lack of an infallible and absolute correlation between knowledge and survival. Some living things get it wrong all of the time and don't survive. All get it wrong some of the time, and some of those do survive. This is why evolutionary epistemology cannot solve the justification problem. But that is a different kind of problem from whether knowledge is possible at all. And the probabilities tell us that living things survive precisely because they are getting enough of it right some of the time. But if Kant were correct, if living things could never know the things-in-themselves, then life would never survive for any time at all beyond that given by chance matching. Since we and the myriad other forms of life do survive, Kant must be wrong. Our a priori intuitions and categories do not make true knowledge impossible, but they also do not make it certainly true always.

And yet, everything in Chapters 5 and 6 tells us that, in positing the existence of a priori intuitions and categories that make experience possible, Kant was, if not absolutely right, very nearly right. We do indeed come into the world with innate determiners of knowledge. These innate determiners, themselves a priori forms of knowledge, are a priori to us as individuals, but they have been gained a posteriori by the long evolutionary history of our species. Many philosophers and scientists have recognized that Kant's categories and intuitions can be understood in evolutionary terms; and hence in this respect he was partly right. He was wrong, of course, in claiming that these innate knowledge forms block us from ever being able to experience the world as it is. But, being more right than wrong, the phrase 'evolutionary Kantianism' is now commonly used to describe the scientific understanding of the existence of a priori or innate psychological devices that make it possible for us to know the world, in the ordinary usage of that word 'know'. Evolutionary epistemology, and specifically the evolutionary epistemology that is built around a multiple-level model of evolution and knowledge, gives us a picture of how such a priori elements are put in place, why they are necessary and why they must give us access to real, if only partial, knowledge.

What, then, of Hume himself and his criticism of induction and causation that so startled Kant and plunged philosophical epistemology into a 250-year-long crisis? Remember, Hume criticized inductive inference because we can never be certain that 'instances of which we have had no experience must resemble those of which we have had experience, and that the course of nature continues always uniformly the same'. If the 'principle of the uniformity of nature' is uncertain, then predictions based upon past experience can never be certain. Hume's point was that the principle of the uniformity of nature is not a logical truth, because we can always imagine a change in the course of nature. And if we have recourse to pleading that, while the principle might not be absolutely correct, at any moment in time it probably holds, then since the probability judgement is itself based upon our past experience, raising such experience, even in probabilistic terms, as a defence against possible future uncertainty is circular and invalid. We can no more trust the projection of past probability into the future than we can trust the inductive inference of anything else. Logically there is no certainty that anything in the future, including probabilities of events occurring or not occurring, will be like the past. We are never justified in drawing inductive inferences. The future is always uncertain.

Now, Hume was, as noted in Chapter 1, correct both logically and factually, in the sense that the world we live in does change all the time. Some change is so slow that the individual observer cannot detect the change, and life is lived as if no change were occurring at all. For example, the continents are indeed drifting slowly, ever so slowly, about the surface of the Earth. In time they change their spatial relationships with one another and a whole series of other changes are a consequence of such movement, including the forms of mountain ranges – the Himalayas, for instance, have been increasing in height for 30–40 million years. But all this is happening so slowly as to be undetectable by ordinary means. (An immediate consequence of continental drift is earthquake, which we certainly do detect; but the relationship between continental drift and earthquakes has only recently become known to science as a culture rather than to any individual, and earthquakes are always reacted to by the individual as unpredictable sudden change.) So we behave and are built as if the continents were stationary. That they do in fact move fools living things constructed on the 'assumption' that they are standing still. The seemingly bizarre migratory behaviour of certain species of animals, like turtles, eels and salmon, is caused by the distortions to their ancient migration routes which result from the shifting land masses. The gene pools of these animals cannot detect such slow change, and so the expressed instinctive behaviour, in remaining the same against a background of constantly changing continental positions, appears extremely odd to the outside observer who sees only the end result.

Other change, as we have seen in Chapter 5, can be much more rapid. So not only is Hume correct, but worse, the uncertainty is not constant. Some things are more uncertain in unit time than others. In a day our social relationships may change, but the state or position of our houses is less likely to do so; and the periodic cycling of long-term weather events such as ice ages, or the drift of continents, occurs so slowly that, to all intents and purposes, the weather and the continents don't change at all. In a decade our houses may indeed alter, in both state and position. If our regular commuting involved journeys of many years rather than minutes or hours, we would be much less surprised to find that our houses were no longer where, or in the state in which, we had left them; and friends might be dead, not just different. But still the weather seems constant. Yet over a 1,000-year period deserts may advance or retreat and temperature levels change significantly, and so too do the spatial distributions of forests and fertile lands change.

In the face of this Humean uncertainty – Humean because modern science can now interpret what the philosopher was pointing to as a factual condition of the universe and not merely a logical point – nature has come up with an elegant and effective response. Unable to rely upon just one level of evolution and one unit of selection as the means of gaining knowledge about just one range of frequencies of change (that range being limited at one end by the change becoming undetectably slow to the point that survival is not threatened by a failure to respond to it, and at the other by the generational deadtime of each species), subsidiary evolutionary processes have evolved, each with its own units of selection, and each able to gain knowledge about changes that are occurring at ever-higher frequencies. So in the real world of biology the logic of Humean uncertainty is converted into a pragmatic issue of dividing the world into band widths of frequencies of change and fluctuation, and employing knowledge-gaining mechanisms that are able to match the rates of perturbation of the world with organic structures able to alter their own states at equivalent rates.

Three and a half billion years of life on Earth tell us that though individuals, individual species and whole larger groupings of living forms may come and go precisely because Humean uncertainty can be literally fatal, the biotic system as a whole endures, being rather adept at solving the problem. Life really is quite good at the knowledge game.

There is one last loose end to tie up. Humean uncertainty, the uncertain futures problem, is just another way of saying that nature is never prescient. Precisely because nature is never prescient it has had to evolve this elaborate, layered structure of knowledge, perched at the top of which is human culture. One part of human culture is science, and yet science really does seem to be prescient, or at least on its way to becoming prescient. We judge the success of science by the accuracy of its prediction of future events and the adequacy of its explanation and depiction of the past. Lunar and solar eclipses can now be predicted for a virtually indefinite future period. The shape of the original continental mass and the forces that tore it apart hundreds of millions of years ago are now understood in great detail. Science is the only level in the whole complex structure that can encompass knowledge of extremely slow change like ice ages and extraordinarily rapid events such as those found in chemical reactions.

Well, then, does science break the dictum that nature is never prescient? Despite the risk of some etymological confusion, perhaps so. What science has become extraordinarily skilled at is providing a metric of Humean scepticism, for quantifying epistemological uncertainty with exquisite accuracy. Science is a kind of Humean uncertainty detector whose success we measure by the extent to which it pins down and cuts back uncertainty – ultimately to the point where it cannot diminish it any further. Physics now tells us that irreducible uncertainty is a characteristic of all physical systems, but we know something of the source and magnitude of this, the ultimate in Humean uncertainty. Such knowledge is an extraordinary achievement, as close to prescience as we will ever come. It is so close that one might want to judge it unnatural knowledge.

Glossary

ADAPTATION: conventionally, some feature or attribute of an organism that helps it survive and reproduce. Adaptations are often thought of as goal- or end-directed, bearing a relationship of fit to some feature of the environment.

BEHAVIOUR: in general usage, behaviour is any doing, acting or even having. Not in this book. I have adopted Piaget's definition of behaviour as adaptive action or doing (and never having, however adaptive the having may be).

CANALIZATION: the evolution of assured developmental pathways by the selection of genes that come eventually to replace initially necessary environmental stimulation. An essential part of genetic assimilation.

CHROMOSOME: a complex thread-like structure containing genetical information arranged in a linear sequence. Chromosomes are found in almost all the cells of the body, even though only parts of some of them are active in each cell. Humans have twenty-three pairs of chromosomes.

COGNITION: the processes and mechanisms by which each individual acquires knowledge (as commonly understood) of their world. May also refer to the products of such processes and mechanisms. Originally one of the three great divisions of psychology, the other two being the conative (willing) and emotive (feeling) spheres.

CONSTRAINT: in ordinary language it points to restraining or restricting forces. In evolutionary theory it refers to the limitations on the variation that can be generated and to limitations on the phenotypic structures that can be built by developmental processes. It is a general principle that not everything goes.

CULTURE: shared knowledge. Traditionally the word is used to describe the totality of thoughts, actions and artefacts of a social group or society, but it often includes a reference to the learning and transmission of such thoughts, actions and artefacts.

DEDUCTION: its widest meaning is a form of analytical reasoning from the general to the less general. In logic, deduction is a form of inference in which the conclusion necessarily follows from the premises.

DNA: short for deoxyribonucleic acid. This is a complex macro-molecule which is the primary carrier of genetic information. Genes are sequences of DNA.

DOMAIN-SPECIFIC INTELLIGENCE: in this book it is argued that there are intelligences rather than some single intelligence, each of which has evolved in response to specific kinds of short-term stabilities. These intelligences may be separate from one another both in terms of the conditions of the world with which they are in adaptive relationship as well as in their embodiment in brain mechanisms and psychological processes.

EMPIRICISM: the school of epistemology in which the central claim to knowledge is that it cannot occur independently of experience. Empiricism entails the denial of universal truths, of the existence of knowledge regardless of experience and of innate knowledge. In its most radical form it holds that all ideas are reducible to sensation.

EPIGENESIS: the doctrine that development is not a simple and inevitable unfolding or growing process, but instead is highly variable within certain limits. The variability results from a cascade of immensely complex interactions between genes, developing features of the organism and the environment in which development occurs.

EPISTEMOLOGY: the branch of philosophy concerned with the methods and the validity of knowledge or knowing. Sometimes defined as the theory of knowledge.

EVOLUTION: transformation in time caused in biological systems by the operation of certain specific processes. To be contrasted with ordinary usage where evolution simply means change in time. Evolution is often defined in terms of its consequences, notably speciation, the formation of adaptations, cumulative change in the characteristics of a population (descent with modification) and changes in gene frequencies in a breeding population. Evolutionary theory is the theory of biological evolution.

EVOLUTIONARY EPISTEMOLOGY: the evolutionary biological study of knowledge. At minimum, the claim is that our ability to know has evolved. However, it is usually taken to have two additional implications. One is that all adaptations are forms of knowledge. The other is that our ability to know has characteristics that are the result of its evolutionary origins. Evolutionary epistemology is treated in this book as science. There are, though, some philosophers who treat evolutionary epistemology as philosophy.

EXAPTATION: an adaptation that either originated as a non-adaptive characteristic or first evolved as an adaptation with a different function from that of the present.

EXOSOMATIC STORAGE: the phrase literally means storage outside of the body. It refers to a set of human artefacts, including books, tapes and disks, by which information can be stored separately from the people who have acquired the knowledge on which the information is based. Some consider exosomatic storage to be a notable and significant feature of human culture. The view taken in this book is generally one of questioning its importance, at least in the evolution of human culture.

FITNESS: a term used by evolutionists in a number of different ways. Most commonly it simply means the ability to survive and reproduce. There are also more technical meanings concerning numbers of offspring (and even the offspring of offspring), as well as the reproductive advantages bestowed by genes or sets of genes.

GENE: the physical unit of heredity. It is usually thought of as a length of DNA that encodes for the amino acid chain that forms a protein. Because of mutation a gene that occupies one place on a chromosome may take different forms, known as *alleles*. Alleles are distinguished from other mutational states of the same gene by the different ways in which they are expressed.

GENE POOL: the totality of genes in a breeding population. An abstract concept, since genes are carried about in separate individuals that may or may not survive and reproduce, thus perpetuating or failing to perpetuate their genes in the gene pool.

GENERATIONAL DEADTIME: the period in sexually reproducing organisms extending from conception to the time when the organism is reproductively competent. It is a lag-time from the moment an organism receives, and can never again change, a sample of genetical information from the gene pool to when the adequacy of that information can be tested by returning or failing to return those genes to the gene pool.

GENETIC ASSIMILATION: the process by which selection leads to genotypes that modify development in a way similar if not identical to that wrought initially by environmental effects. This is an empirically well-demonstrated effect.

GENETIC EPISTEMOLOGY: the variant of evolutionary epistemology owed to Piaget which focuses primarily on human cognition. It differs from other forms of evolutionary epistemology in that the central role of chance in evolution is rejected, specifically the notion that chance variation is the ultimate source both of new genetic material and of its dissemination through breeding populations. Genetic epistemology is built around the structuralist ideas of organization, self-regulation, co-ordination and construction as the basis for transformation in time rather than the Darwinian concepts of variation and selection.

GENOTYPE: its proper meaning refers to genetic constitution at a specific place on the chromosomes. But genotype is widely, loosely and inaccurately used to mean the entire genetic complement of an organism. Synonymous with *genome*.

GRADUALISM: the doctrine that evolutionary change is gradual rather than occurring in leaps or jumps or jerks. It takes in the view that neither new species nor complex adaptations in individuals leap fully formed into the world.

HEURISTIC: in psychology and cognitive science at large it refers to a simple procedure or strategy for solving a problem, or moving towards a solution of a problem, the procedure usually having some generality of use. In ordinary language a heuristic is that which leads to discovery and invention. In this book it is employed in a manner closer to ordinary usage in that a heuristic is considered to be a set of processes by which adaptations are invented or discovered. These processes are those by which variants are generated, selected and then regenerated and hence propagated. A hierarchy of such g-t-r (generate–test–regenerate) heuristics is described. The *primary heuristic* is made up of g-t-r mechanisms operating at the genetic and developmental level, which is the most fundamental. The *secondary heuristic* comprises g-t-r mechanisms found within the immune system and brain which have evolved in response to limitations on the ability of the primary heuristic to form adaptations to certain features of the world. The *tertiary heuristic* is made up of the g-t-r mechanisms of culture.

HIERARCHY: a type of ordering of entities. There are two fundamental forms. A *structural hierarchy* is characterized by physical containment. A *control hierarchy* is characterized by a scale of authority and control.

IMMUNE SYSTEM: a complex organ system that protects the body from foreign substances that enter it, especially invading microorganisms or parts of organisms. In this book the immune system is considered to be part of the secondary heuristic.

INDUCTION: a form of reasoning by which a number of observed particularities or facts become the basis for a general assertion. Induction is thus a form of ampliative reasoning by which an inference is made about all the members of a class from observation of only some members of that class.

INNATE: existing at birth. Hence inborn or congenital.

INSTINCT: a species-typical adaptive behaviour which is not affected or modified by the operation of the secondary heuristic as defined in this book.

INSTRUCTIONISM: synonymous with Lamarckianism, where adaptive traits are produced in response to environmental conditions. In this case, evolution is directed, the changes that occur being instructed or impressed upon organisms. In instructionist evolutionary change the environment is the ruling causal force.

INTELLIGENCE: normally refers to a quality of human mind or, less commonly, to the excellence of a behavioural adaptation. In this book it is used synonymously with rationality and refers to the secondary heuristic based in the brain.

INTERACTORS: entities whose interaction with the environment results in differential replicator survival. Individual organisms are paradigmatic interactors.

JUSTIFICATION: a term of appraisal as to the reasonableness of a claim to knowledge. Justification is what distinguishes a belief that is knowledge from one that is not knowledge.

LEARNING: adaptive behavioural change that is due to the operation of the secondary heuristic. Conventionally, learning is any long-lasting modification in behaviour due to experience, but this confuses the primary (especially the developmental component thereof) and secondary heuristics.

LINEAGES: these are the entities that change indefinitely in time as a result of replication and interaction.

LONG-TERM STABILITY: refers to a period relative to the average life span of individuals within a species when either the values of parts of the world that are important to those organisms are constant or their rates of alterations within fixed limits are such that these conditions can be adapted to entirely by the primary heuristic of genetics and development.

LYMPHOCYTES: these are cells of the immune system that play a crucial role in defence against foreign substances. Some of them are responsible for the production of antibodies.

MACROEVOLUTION: large-scale changes in populations and species that result in speciation. Often refers to fragmentation and change at or beyond the species level.

MEME: the unit of cultural heredity analogous to the gene.

MICROEVOLUTION: evolutionary changes that occur within a species or even a breeding population. Usually measured by changes in gene frequencies and relatively small phenotypic effects, microevolution is held by Darwinists and neo-Darwinists to be the driving force of macroevolution.

MODULARITY: the doctrine that the construction of the mind is modular, sometimes discussed in terms of the vertical organization of psychological faculties or functions as opposed to a horizontal mental architecture. Also sometimes referred to in terms of domain-specificity of mechanism and psychological function.

MOLECULAR DRIVE: the notion that events occurring at the molecular level may give rise to genetic change unrelated to natural selection effects. It is not an anti-Darwinian stance, since it does not deny the existence and importance of selection.

MUTATION: the ultimate source of genetical variation and the cause of allelic differences in genes, it is a change in DNA sequence or chromosome structure.

NATURAL SELECTION: Darwin's great discovery, often abbreviated to selection. A highly abstract concept, it means the differential survival and/or production of organisms that is non-random and leads to the differential propagation of genes. Natural selection occurs under natural or 'wild' conditions, as opposed to the artificial selection imposed by plant and animal breeders.

NEO-DARWINISM: refers to the fusion of Darwinian natural selection and Mendelian genetics achieved in this century, sometimes known as the *modern synthesis* or the *synthetic theory of evolution*. The emphasis is on the selection of chance variation as the basis of adaptation, and the necessary relationship between microevolution and macroevolution.

NON-EQUILIBRIUM THERMODYNAMIC THEORY OF EVOLUTION: very much a minority approach to evolution, it is couched in the language of thermodynamic theory. Entropy (disorder), energy and information are key terms and concepts, with natural selection reduced in importance to a determiner of rates of evolutionary change. It is a physical science eye-view of evolution.

ONTOGENY: a technical word for individual development.

ONTOLOGY: the branch of philosophy concerned with what exists, quite apart from whether we can know it or not. Sometimes defined as the theory of being.

ORGANIC SELECTION: a vague term with a number of different meanings ranging from the dangerously Lamarckian to near-common sense. At its most general it is a hypothesis about how phenotypic plasticity may facilitate reconstruction of the genotype. Baldwin, the originator of the idea, favoured an interpretation in which variation was conceived as being both genetically and developmentally caused, with the two acting in concert, possibly resulting in selection for developmental pathways that, though initially largely environmentally determined, become solely genetically determined.

PHENOTYPE: the organism in flesh-and-blood form. The expression of genetical information via the processes of development, to result in, for example, you or me.

PREDICTABLE UNPREDICTABILITY: a form of change that has been central to the evolution of intelligence or rationality. It describes a situation where events in the world fluctuate in some predictable way, within certain limits, so that the primary heuristic can gain knowledge of such predictable change and evolve adaptations for dealing with it. But superimposed upon this fundamental pattern of change is a degree of unpredictability, which means that the precise values of events can only be crudely predicted. The secondary heuristic evolves in order to acquire a more accurate fix on the values of these unpredictable elements. It is a very important conception for this book, because it is the basis for the claim that intelligence is never a *tabula rasa*, but always nested under and primed by the primary heuristic of genes and development.

PUNCTUATED EQUILIBRIUM THEORY: a theory of evolution, put forward in the 1970s by Eldredge and Gould, which claims that evolution is not gradual, as Darwin had insisted, but comprises instead a pattern of little or no change – stasis – for long periods of time, interrupted by geologically brief periods of rapid change – punctuation – during which time speciation occurs. The most radical feature of the theory is its denial of the causal links between microevolution and macroevolution. Less controversially, the theory postulates a hierarchy of selection processes with different units of selection at each level.

RATIONALISM: the school of epistemology in which the criterion of true knowledge is intellectual and deductive, rather than sensory and experiential. According to rationalists, knowledge is the product of pure reason and not of sense perception.

RATIONALITY: as used in this book, it is synonymous with intelligence, which is a product of the secondary heuristic. In general parlance it merely refers to an ability to think and reason.

REINFORCEMENT: a technical term from learning theory. A reinforcement is any event that alters the probability of a response in a way that is not due to chance.

REPLICATOR: any entity of which copies can be made. The term was adopted as a way of writing a general theory of evolution, and used in this book as a way of distinguishing the different mechanisms underlying different units of selection. Genes are paradigmatic replicators.

SATISFICING: the adoption not of the best or optimal solution to a problem but of one that is good and satisfactory. A kind of 'it will do' solution in contrast to an 'it is the best' solution.

SELECTIONISM: another word for Darwinism. The central feature of selectionism is that the traits that evolve initially occur independently of their potential usefulness in the sense of their contribution to fitness. In other words, selectionism refers to blind and undirected processes as the causes of evolution. In selectionist evolutionary change, therefore, the generators of variant states are co-equal causes of evolution along with the environmental conditions that are the source of selection forces.

SHORT-TERM STABILITY: refers to the pattern of change in the unpredictable element of predictable unpredictability. It is a period of time of relative stability in the value of events in the world that are significant for those organisms. The length of this period will vary relative to the average life span of the individuals of a species. To be contrasted with *instability* of the unpredictable element, where fluctuations are so rapid relative to life span that it is not worth storing their values.

STRUCTURALISM: schools of thought in psychology, anthropology, linguistics, literary analysis and evolutionary biology which share the notion that structures are to be understood in terms of wholeness, self-regulation and transformation.

TABULA RASA: means a blank slate or tablet. John Locke argued ——— that the mind is at birth like a blank slate upon which experience writes, and hence denied the existence of innate, or a priori, knowledge.

TELEOLOGY: the belief that final causes exist which shape prior events. Hence it is a doctrine that is contrary to normal scientific views that causes occur before their effects. Sometimes used in a general sense to refer to design and purpose in nature.

UNCERTAIN FUTURES PROBLEM: because nature is never prescient, every organism enters the world without any certainty that events in its future will be like those in the past that formed the selection pressures which led to the evolution of the lineage of which it is a part. In this book, the uncertain futures problem is considered to be the reason for the evolution of the secondary heuristic, which includes learning and intelligence.

UNIT OF SELECTION: one of the central theoretical issues in evolutionary theory. The unit of selection is what adaptations are 'good for', that is, what is preserved and propagated in time. Not to be confused with what selection acts on or what the units of variation may be, though the latter are often more closely identified with the units of selection than are the former. Darwin thought the individual to be the unit of selection; some biologists, albeit a very small number, believe that groups might be a unit of selection; punctuationists incline to the view that the species may be a unit of selection; and some evolutionary epistemologists, like the author, believe that antibodies in the immune system and memories in the brain are also units of selection. One of the arguments pursued in this book is that most of these might be units of selection at one and the same time; that evolutionary theory must be extended to a hierarchical structure with the same evolutionary processes operating through different mechanisms at each level; and that each level has its own unit of selection.

UNIVERSAL DARWINISM: the idea that Darwin's principles of evolution are fundamental to all life everywhere in the universe.

In this book the phrase is used to mean that Darwinian evolution here on Earth is not confined to a between-organism set of processes that results in adaptation and speciation, but is also found within organisms, where these same processes, embodied in the mechanisms of the immune and central nervous systems, are responsible for some of the transformations in time of these systems.

Bibliography

CHAPTER 2

Bock, W.J. (1980) 'The definition and recognition of biological adaptation', in *American Zoologist*, 20: 217–27.

Campbell, J.H. (1985) 'An organizational interpretation of evolution', in D.J. Depew and B.H. Weber (eds.), *Evolution at a Crossroads: The New Biology and the New Philosophy*, Cambridge, Mass.: MIT Press.

Darwin, C. (1859) *On the Origin of Species by Natural Selection or the Preservation of Favoured Races in the Struggle for Life*, London: Murray.

Darwin, C. (1974) *Autobiography*, Oxford: Oxford University Press.

Dawkins, R. (1976) 'Hierarchical organization: a candidate principle for ethology', in P.P.G. Bateson and R.A. Hinde (eds.), *Growing Points in Ethology*, Cambridge: Cambridge University Press.

Desmond, A. and Moore, J. (1991) *Darwin*, London: Michael Joseph.

Dover, G. (1982) 'Molecular drive: a cohesive mode of species evolution', in *Nature*, 299: 111–17.

Eldredge, N. and Gould, S.J. (1972) 'Punctuated equilibria: an alternative to phyletic gradualism', in T.J.M. Schopf (ed.), *Models in Paleobiology*, San Francisco: Freeman.

Eldredge, N. and Salthe, S. (1984) 'Hierarchy and evolution', in *Oxford Surveys in Evolutionary Biology*, 1: 184–208.

Fisher, R.A. (1930) *The Genetical Theory of Natural Selection*, Oxford: Clarendon Press.

Gould, S.J. (1982) 'The meaning of punctuated equilibrium and its role in validating a hierarchical approach to macroevolution', in R. Milkman (ed.), *Perspectives on Evolution*, Sunderland, Mass.: Sinauer.

Gould, S.J. and Lewontin, R.C. (1979) 'The spandrels of San Marco and the Panglossian paradigm: a critique of the adaptationist programme', in *Proceedings of the Royal Society of London*, Series B, 205: 581–98.

Haldane, J.B.S. (1932) *The Causes of Evolution*, New York: Longmans.

Hempel, C.G. (1965) *Aspects of Scientific Explanation*, New York: The Free Press.

Kimura, M. (1983) *The Neutral Theory of Molecular Evolution*, Cambridge: Cambridge University Press.

King, J.L. and Jukes, T.H. (1969) 'Non-Darwinian Evolution', in *Science*, 164: 788–98.

Lewontin, R.C. (1981). 'On constraints and adaptation' in *Behavioral and Brain Sciences*, 4: 244–5.

Lewontin, R. C. (1983) 'Gene, organism and environment', in D.S. Bendall (ed.), *Evolution from Molecules to Men*, Cambridge: Cambridge University Press.

Maynard-Smith, J. (1989) *Evolutionary Genetics*, New York: Oxford University Press.

Mayr, E. (1976) *Evolution and the Diversity of Life*, Cambridge, Mass.: Harvard University Press.

Mayr, E. (1982) *The Growth of Biological Thought*, Cambridge, Mass.: Harvard University Press.

Pattee, H.H. (1970) 'The problem of biological hierarchy', in C.H. Waddington (ed.), *Towards a Theoretical Biology*, Vol. 3: *Drafts*, Edinburgh: Edinburgh University Press.

Pattee, H.H. (1973) (ed.) *Hierarchy Theory*, New York: Braziller.

Simon, H.A. (1982) *The Sciences of the Artificial*, 2nd edn, Cambridge, Mass.: MIT Press.

Sommerhoff, G. (1950) *Analytical Biology*, Oxford: Oxford University Press.

Williams, G.C. (1966) *Adaptation and Natural Selection*, Princeton: Princeton University Press.

Williams, G. C. (1992) Natural Selection: *Domains, Levels and Challenges*, New York: Oxford University Press.

Wright, S. (1930), 'Evolution in Mendelian populations', in *Genetics*, 16: 97–159.

CHAPTER 3

Ada, G.L. and Nossal, G. (1987) 'The clonal-selection theory', in *Scientific American*, 257(2): 50–57.

Baldwin, J.M. (1909) *Darwin and the Humanities*, Baltimore: Review Publishing Co.

Burnet, F.M. (1959) *The Clonal Selection Theory of Acquired Immunity*, London: Cambridge University Press.

Campbell, D.T. (1960) 'Blind variation and selective retention in creative thought as in other knowledge processes', in *Psychological Review*, 67: 380–400.

Cziko, G.A. and Campbell, D.T. (1990) 'Comprehensive evolutionary epistemology bibliography', in *Journal of Social and Biological Structures*, 13: 41–82.

Darwin, C. (1871) *The Descent of Man*, London: Murray.

Dawkins, R. (1976) *The Selfish Gene*, Oxford: Oxford University Press.

Dawkins, R. (1982) 'Replicators and vehicles', in King's College Sociobiology Group's *Current Problems in Sociobiology*, Cambridge: Cambridge University Press.

Hamilton, W.D. (1964) 'The genetical evolution of social behaviour', I & II, in *Journal of Theoretical Biology*, 7: 1–16 and 17–51.

Hardcastle, V.G. (1993) 'Evolutionary epistemology as an overlapping, interlevel theory', in *Biology and Philosophy*, 8: 173–92.

Hull, D.L. (1980) 'Individuality and selection', in *Annual Review of Ecology and Systematics*, 11: 311–32.

Hull, D.L. (1988a) 'Interactors versus vehicles', in H.C. Plotkin (ed.), *The Role of Behaviour in Evolution*, Cambridge, Mass.: MIT Press.

Hull, D.L. (1988b) *Science as Process*, Chicago: Chicago University Press.

James, W. (1880) 'Great men, great thoughts, and the environment', in *The Atlantic Monthly*, 46(276): 441–59.

Jerne, N.K. (1967) 'Antibodies and learning: selection versus instruction', in G.C. Quarton, T. Melnechuk and F.O. Schmitt (eds.), *The Neurosciences: A Study Program*, New York: Rockefeller University Press.

Jerne, N.K. (1985) 'The generative grammar of the immune system', in *Science*, 229: 1057–9.

Lewontin, R.C. (1970) 'The units of selection', in *Annual Review of Ecology and Systematics*, 1: 1–18.

Piaget, J. (1980) *Adaptation and Intelligence: Organic Selection and Phenocopy*, Chicago: University of Chicago Press.

Plotkin, H.C. (1991) 'The testing of evolutionary epistemology', in *Biology and Philosophy*, 6: 481–97.

Popper, K.R. (1972) *Objective Knowledge: An Evolutionary Approach*, Oxford: Oxford University Press.

Schull, J.(1990) 'Are species intelligent?', in *Behavioral and Brain Sciences*, 13: 63–108.

Waddington, C.H. (1952) 'Selection of the genetic basis for an acquired character', in *Nature*, 169: 278–9.

Waddington, C.H. (1969) 'Paradigm for an evolutionary process', in C.H. Waddington (ed.), Towards a Theoretical Biology, Vol. 2: Sketches, Edinburgh: Edinburgh University Press.

Wynne-Edwards, V.C. (1962) Animal Dispersion in Relation to Social Behaviour, Edinburgh: Oliver & Boyd.

CHAPTER 4

Boakes, R. (1984) From Darwin to Behaviourism: Psychology and the Minds of Animals, Cambridge: Cambridge University Press.

Cullen, E. (1957) 'Adaptations in the kittiwake to cliff-nesting', in Ibis, 99: 275–302.

Darwin, C. (1859) On the Origin of Species by Means of Natural Selection or the Preservation of Favoured Races in the Struggle for Life, London: Murray.

Darwin, C. (1871) The Descent of Man, London: Murray.

Frisch, K. von (1954) The Dancing Bees, London: Methuen.

Gould, J.L. (1990) 'Honey bee cognition', in Cognition, 37: 83–103.

Harlow, H.F. (1962) 'The heterosexual affectional system in monkeys', in American Psychologist, 17: 1–9.

Harlow, H.F. and Harlow, M.K. (1965) 'Social deprivation in monkeys', in Scientific American, 207: 136–46.

Johnston, T.D. (1982) 'Selective costs and benefits in the evolution of learning', in Advances in the Study of Behaviour, 12: 65–106.

Lehrman, D.S. (1953) 'A critique of Konrad Lorenz's theory of instinctive behavior', in Quarterly Review of Biology, 28: 337–63.

Lorenz, K.Z. (1950) 'The comparative method in studying innate behaviour patterns', in Symposia of the Society for Experimental Biology iv, Physiological Mechanisms in Animal Behaviour: 221–68.

Lorenz, K.Z. (1965) Evolution and Modification of Behaviour, Chicago: Chicago University Press.

Mayr, E. (1961) 'Cause and effect in biology', in Science, 134: 1501–6.

Oppenheim, R.W. (1982) 'Preformation and epigenesis in the origins of the nervous system and behaviour: Issues, concepts and their history', in P.P.G. Bateson and P.H. Klopfer (eds.), Perspectives in Ethology, Vol. 2: Ontogeny, New York: Plenum.

Piaget, J. (1979) Behaviour and Evolution, London: Routledge & Kegan Paul.

Schneirla, T.C. (1953) 'Basic problems in the nature of insect behavior', in K.D. Roeder (ed.), *Insect Physiology*, New York: Wiley.

Tinbergen, N. (1959) 'Comparative studies of the behaviour of gulls (Laridae)', in *Behaviour*, 15: 1–70.

Wilson, E.O. (1971) *The Insect Societies*, Cambridge, Mass.: The Belknap Press of Harvard University Press.

CHAPTER 5

Barlow, G.W. (1991) 'Nature–nurture and the debates surrounding ethology and sociobiology', in *American Zoologist*, 31: 286–96.

Campbell, D.T. (1974) 'Evolutionary epistemology', in P.A. Schilpp (ed.), *The Philosophy of Karl R. Popper*, La Salle, Illinois: Open Court.

Changeux, J.P. (1983) *L'homme Neuronal*, Paris: Fayard.

Changeux, J.P. and Dehaene, S. (1989) 'Neuronal models of cognitive functions', in *Cognition*, 33: 63–109.

Corning, W.C. and Kelly, S. (1973) 'Platyhelminthes: the Turbellarians', in W.C. Corning, J.A. Dyal and A.O.D. Willows (eds.), *Invertebrate Learning*, Vol. 1: *Protozoans Through Annelids*, New York: Plenum.

Dyal, J.A. (1973) 'Behaviour modification in Annelids', in W.C. Corning, J.A. Dyal and A.O.D. Willows (eds.), ibid.

Edelman, G. (1987) *Neural Darwinism: The Theory of Neuronal Group Selection*, New York: Basic Books.

Hinde, R.A. and Stevenson-Hinde, J. (eds.) (1973) *Constraints on Learning: Limitations and Predispositions*, London: Academic Press.

Kroodsma, D.E. (1988) 'Contrasting styles of song development and their consequences among passerine birds', in R.C. Bolles and M.D. Beecher (eds.), *Evolution and Learning*, Hillsdale, New Jersey: Erlbaum.

Lorenz, K.Z. (1965) *Evolution and Modification of Behaviour*, Chicago: Chicago University Press.

MacArthur, R.H. and Wilson, E.O. (1967) *The Theory of Island Biogeography*, Princeton: Princeton University Press.

Oyama, S. (1985) *The Ontogeny of Information: Developmental Systems and Evolution*, Cambridge: Cambridge University Press.

Piaget, J. (1953) *The Origin of Intelligence in the Child*, London: Routledge & Kegan Paul.

Plotkin, H.C. (1988a) 'Evolutionary epistemology as science', in *Biology and Philosophy*, 2: 295–313.

Plotkin, H.C. (1988b) 'Behaviour and evolution', in H.C. Plotkin (ed.), *The Role of Behaviour in Evolution*, Cambridge, Mass.: MIT Press.

Plotkin, H.C. and Odling-Smee, F.J. (1979) 'Learning, change and evolution: an enquiry into the teleonomy of learning', in *Advances in the Study of Behaviour*, 10: 1–41.

Plotkin, H.C. and Odling-Smee, F.J. (1982) 'Learning in the context of a hierarchy of knowledge gaining processes', in H.C. Plotkin (ed.), *Learning, Development and Culture: Essays in Evolutionary Epistemology*, Chichester: Wiley.

Waddington, C.H. (1969) 'Paradigm for an evolutionary process', in C.H. Waddington (ed.), *Towards a Theoretical Biology*, Vol. 2: *Sketches*, Edinburgh: Edinburgh University Press.

CHAPTER 6

Bateson, G. (1968) 'Redundancy and coding', in T.A. Sebeok (ed.), *Animal Communication*, Bloomington: Indiana University Press.

Bruce, V. (1988) *Recognising Faces*, Hove: Erlbaum.

Cheng, P.W. and Holyoak, K.J. (1989) 'On the natural selection of reasoning theories', in *Cognition*, 33: 285–313.

Chomsky, N. (1980) *Rules and Representations*, New York: Columbia University Press.

Chomsky, N. (1988) *Language and Problems of Knowledge*, Cambridge, Mass.: MIT Press.

Cosmides, L. (1989) 'The logic of social exchange: Has natural selection shaped how humans reason? Studies with the Wason selection task', in *Cognition*, 31: 187–276.

Dawkins, R. (1976) *The Selfish Gene*, Oxford: Oxford University Press.

Donald, M. (1991) *Origins of the Modern Mind*, Cambridge, Mass.: Harvard University Press.

Gigerenzer, G. and Hug, K. (1992) 'Domain-specific reasoning: social contracts, cheating, and perspective change', in *Cognition*, 43: 127–71.

Heyes, C.M. and Plotkin, H.C. (1989) 'Replicators and interactors in cultural evolution', in M. Ruse (ed.), *What the Philosophy of Biology Is*, Dordrecht: Kluwer.

Humphrey, N.K. (1976) 'The social function of intellect', in P.P.G. Bateson and R.A. Hinde (eds.), *Growing Points in Ethology*, Cambridge: Cambridge University Press.

Johnson, M.H. and Morton, J. (1991) *Biology and Cognitive Development: The Case of Face Recognition*, Oxford: Blackwell.

Kahneman, D., Slovic, P. and Tversky, A. (eds.) (1982) *Judgement under Uncertainty: Heuristics and Biases*, Cambridge: Cambridge University Press.

Karmiloff-Smith, A. (1992) *Beyond Modularity: A Developmental Perspective on Cognitive Science*, Cambridge, Mass.: MIT Press.

Laland, K.N. (1993). 'The mathematical modelling of human culture and its implications for psychology and the human sciences', in *British Journal of Psychology*, 84: 145–69.

Lenneberg, E.H. (1967) *Biological Foundations of Language*, New York: Wiley.

Meier, R.P. (1991) 'Language acquisition by deaf children', in *American Scientist*, 79: 60–70.

Pinker, S. and Bloom, P. (1990) 'Natural language and natural selection', in *Behavioral and Brain Sciences*, 13: 707–84.

Plotkin, H.C. (1988) 'Behaviour and evolution', in H.C. Plotkin (ed.), *The Role of Behaviour in Evolution*, Cambridge, Mass.: MIT Press.

Pollard, P. (1990). 'Natural selection for the selection task: Limits to social exchange theory', in *Cognition*, 36: 195–204.

Savage-Rumbaugh, E.S. (1986) *Ape Language: from Conditioned Response to Symbol*, Oxford: Oxford University Press.

Schank, R.C. (1982) *Dynamic Memory*, Cambridge: Cambridge University Press.

Seyfarth, R.M. and Cheney, D.L. (1992) 'Meaning and mind in monkeys', in *Scientific American*, 267: 78–84.

Tversky, A. and Kahneman, D. (1973) 'Availability: A heuristic for judging frequency and probability', in *Cognitive Psychology*, 5: 207–32.

Wason, P.C. (1966) 'Reasoning', in B.M. Foss (ed.), *New Horizons in Psychology*, Vol. 1, Harmondsworth: Penguin.

Wason, P.C. (1983) 'Realism and rationality in the selection task', in J. Evans (ed.), *Thinking and Reasoning: Psychological Approaches*, London: Routledge & Kegan Paul.

Index

Names in **bold** refer to first authors in the suggested reading lists at the end of each chapter or to first authors in the Bibliography. These names may also appear in the text. Items marked with an asterisk (★) are listed in the Glossary.